Sungho Kim, In So Kweon

I0014648

Object Identification and Categorization with Visual Context

Sungho Kim, In So Kweon

Object Identification and Categorization with Visual Context

Hierarchical Graphical Model-based Approaches

VDM Verlag Dr. Müller

Imprint

Bibliographic information by the German National Library: The German National Library lists this publication at the German National Bibliography; detailed bibliographic information is available on the Internet at http://dnb.d-nb.de.

Cover image: www.purestockx.com

Publisher:
VDM Verlag Dr. Müller Aktiengesellschaft & Co. KG, Dudweiler Landstr. 125 a, 66123 Saarbrücken, Germany,
Phone +49 681 9100-698, Fax +49 681 9100-988,
Email: info@vdm-verlag.de

Produced in USA and UK by:
Lightning Source Inc., La Vergne, Tennessee, USA
Lightning Source UK Ltd., Milton Keynes, UK
BookSurge LLC, 5341 Dorchester Road, Suite 16, North Charleston, SC 29418, USA

ISBN: 978-3-639-01033-6

Contents

List of Tables

List of Figures

CHAPTER I
INTRODUCTION

1.1 Motivation and goal of work

Vision has very important roles in human life, since it provides over 90% of the information that our brain receives from the external world. The main goal of the vision is to interpret and to interact with the environments where we live in. The core function of the visual interpretation and interaction is recognizing objects around us. Humans can recognize or identify thousands of objects such as human faces, cars, traffic signs, food and so on effortlessly. This is not due to the simplicity of the recognition tasks but due to the excellent visual system.

Recently, several humanoid robots such as HUBO in KAIST, MARU/ARA in KIST and ASIMO in HONDA, have been developed. How can we provide these robots the function of human like vision? Although a great number of fundamental problems such as edge detection, stereo matching, and geometric interpretation have been solved over last half century, the high-level interpretation problems are far ahead of us. The currently available computer vision solutions by far underlay the human vision regarding robustness and performance. Humans are rarely aware of the changes in an object's appearance occurred by changes in viewing direction or illuminations. At the same time, human can also readily group instance of objects and still discriminate on sub-categorical levels. However, even the smallest change in visual sensing environment provided to a computer can make all the differences. This makes it difficult to develop human-like vision in computer vision.

Based on such motivations, the goal is to recognize objects in real world as shown in Fig. 1.1. The definition of object recognition is to infer object identity with its position. In this work, we discuss issues generated by real environment and propose robust methods.

Figure 1.1. The goal is to infer object identities and positions.

1.2   Scalability and generality

The major problems in object recognition in real world are **scalability** and **generality**. The scalability issue is how to recognize a lot of objects successfully in real world. It is related to discriminability or selectivity. In general, as the number of objects increases, the performance of object recognition systems tend to be degraded due to several sources of visual ambiguity. Camera noise or motion blur will remove details of object information as shown in Fig. 1.2. We need a systematic strategy for the disambiguation or discrimination. There is memory size and computational complexity for scalable object recognition.

The generality is another important issue in object recognition. It is related to robustness or invariance to unexpected visual variations. An object recognition system is able to cope with unlearned (novel) environment with finite number of training images. The unlearned environment can be different illumination or different viewpoint for learned objects. Different background with novel objects (not seen) is the hardest case for generality as shown in Fig. 1.2 (right). In this work, we focus on disambiguation for scalability and novel object categorization in cluttered background for generality. Related works to these issues will be introduced in the next chapter and each specific chapter.

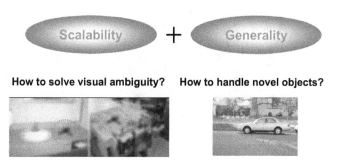

Figure 1.2. Scalability and generality are the major issues in real world object recognition.

1.3 Scope of the work

Object identification: In this work, the main objective is to identify objects in real environment. We don't know the illumination conditions, viewpoints to objects. There are several objects with background clutter. Input information can be static image or video. The number of objects in DB is high (usually over 100).

Object categorization: It is very dangerous to assume that there are always known objects in real world. There can be different objects not seen during learning. As we said, this is one of the biggest generalization issue. We can handle it by object categorization. Simultaneously categorizing and figure-ground segmenting novel objects in real world is most challenging problem.

Visual context and cooperative identification and categorization: First, we focus on visual context to solve ambiguities in object identification and categorization (see Fig. 1.3). Visual context defines visual relations among visual components and provides prediction and interaction. We categorize it spatial context, hierarchical context, and temporal context according to the level of context. Different level of context shows different disambiguation capabilities. Details are explained in Chapter 2. Second, we find that object identification and categorization are cooperative not completely independent. Previous approaches tried to solve independently. Before 2002, most researches focused on object identification. Currently, object categorization is hot research topic in vision community. Examples used to object identification can provide useful similarity information to categorizing novel objects. Conversely, object categorization can update novel object examples in identification and reduce the search space.

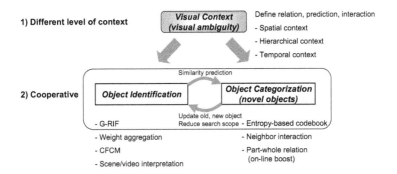

Figure 1.3. Scope of the work for the scalability and generality in object recognition.

1.4   The contributions

This work deals with the problem of object identification in real environment (fully uncontrolled). The main issues of scalability and generality is achieved by utilizing visual context and object categorization. The contributions of this work are summarized as follows.

1. Issues generated by real world (uncontrolled environment) are handled.

   In real world, we don't know object pose, illumination condition, background, object type (seen, novel). We analyze those problems and propose robust methods throughout this work.

2. We analyze the different level of context systematically.

   Several types of visual context exist in real world. In this work, we summarize them as spatial context, hierarchical context and temporal context according to the aspect of visual relation. Different contexts exist in different environments and we propose suitable modeling methods.

3. We show the feasibility of graphical model-based modeling for visual context.

   We propose hierarchical graphical model-based methods for the integration of visual context in object identification and categorization problem. Feasibilities are validated through large scale experiments in real environment.

4. Cooperative object identification and categorization is introduced.

   We find that object identification and categorization are not separate problem but cooperative. Instances in object identification provide similarity to object categorization. Categorization provides generality to identification.

5. Integration of scene, object, part context is proposed and validated by robust scene interpretation.

   After partial modeling of visual context and analyzing property, we propose an integration of whole context in general environment.

6. Simultaneous object categorization and figure-ground segmentation is proposed using the part-part and part-object context.

   Generality for novel object is handled by the examplar-based object categorization. Part-part and part-object context are modeled by boosted MCMC method in Bayesian net. This can categorize novel objects and segment figural region robustly.

1.5 Organization

Background related this work is introduced in Chapter 2. The history and trend of previous works are summarized. The concept of visual context and roles are explained with examples around us. Basics of graphical model is briefly introduced. In part I, we handle the object identification problem according to the level of context. In Chapter 3, we present a biologically motivated feature called G-RIF (Generalized Robust Invariant Feature). Pixel context is utilized to encode local visual information. Chapter 4 shows a background robust object labeling using part context. We model the Gestalt's law of similarity and proximity by weight-aggregation. Chapter 5 handles scalability issues in part-object context based object identification and localization. Chapter 6 integrate the whole static visual context using a hierarchical graphical model for scene interpretation. This is extended to video interpretation by additionally incorporating temporal context in Chapter 7. Please see the Fig. 1.4 for the book organization. In part II, we handle object categorization problem. In Chapter 8, we introduce a entropy-based codebook generation which is robust to surface markings. In Chapter 9, we propose a simultaneous object categorization and figure-ground segmentation method by utilizing part context and part-object context with a directed graphical model (Bayesian Net). Object categorization and

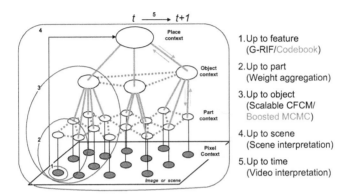

Figure 1.4. Oragnization of the book.

segmentation is done using the proposed boosted MCMC (Markov Chain Monte Carlo). We conclude the work and discuss future work in Chapter 10.

Chapter II
Background and Basics

In this chapter, we introduce the background such as the history of object recognition and types of visual context and their properties in object recognition. Finally, we present basics of graphical model which are useful to model visual context.

2.1 History of object recognition

A lot of researchers in computer vision, cognitive psychology, and neurophysiology society have been studied on the mechanism of human visual object recognition. Especially, neurophysiologists find neuronal mechanisms of visual perception using fMRI, lesion, or spike timing [106]. Cognitive psychologists characterize the mechanisms underlying human cognition in the domains of visual, auditory perception to problem solving, reasoning, learning and memory using eye movement monitoring, behavioral methods. Computer vision scientists develop computational methods based on these models. In this section, we first introduce several key theories of object recognition and related computational approaches with pros and cons. Then, we briefly introduce the state-of-the art object recognition methods and point out the status of solved problems. Finally, we introduce several related works with this book.

2.1.1 Object recognition theory

The history of object recognition theory can be summarized as shown in Fig. 2.1. The theories explain how human visual systems represent objects balancing selectivity (scalability) and invariance (generality). Marr and Nishihara represented an object as a set of primitives which are composed of generalized cones [59]. The scheme is completely view-point independent and became the basis of model-based object recognition. In late 1980s, Biederman proposed RBC (recognition by component) theory [8]. This theory explains that an object is composed of parts (called

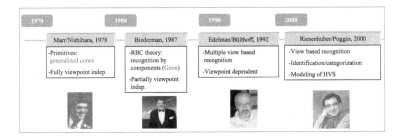

Figure 2.1. Four influential theories of object recognition.

Geons) and their relations. Object representation in this theory is partially view-point independent and basis of the part-based recognition paradigms.

But, Marr and Biederman's models have some problems. First, the models group objects from different classes into the same class (for example, horse and cow). Second, the models group objects from the same class into different classes (for example, round vs. square table). Finally, 3D object recognition is often not view-point independent. In early 1990s, Edelman and Bülthoff experimentally showed that object recognition is view-point dependent [17]. The error rate of 3D object recognition according to view-point is not flat as would be predicted by a view-invariant theory. Edelman and Bülthoff's view-dependent theory is supported by the psychophysical, physiological evidence and suitable for identification problem. Currently, this theory is the basis to the appearance or view-based object recognition approaches. Recently, Riesenhuber and Poggio proposed a hierarchical view-based recognition method (or HMAX model) by approximately modeling the role of simple cell and complex cell of human visual receptive field [80]. This model is based on view-dependent but further composed of tuning by simple cell and max selection by complex cell which gives selectivity and invariance respectively. Currently, the HMAX model is most plausible computational model of human's object recognition.

2.1.2   Previous computational models

Based on the object recognition theories, a lot of different computational methods were proposed during last 30 years. Object representation, learning and recognition methods are different for difference recognition levels (categorization, identifica-

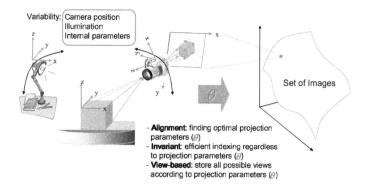

Figure 2.2. Variation compensation schemes.

tion) or object types (2D/3D object, rigid/free form, without texture/with texture) to recognize. Representation methods can be categorized in terms of description level, description scope, description coordinates, object segmentation, feature dimension and invariance. All these aspects concern how to minimize the recognition problems mentioned in previous section.

Researches during 1980s - late 1990s focused on the development of model-based object recognition system. The target objects are usually textureless polygons so geometric information is usually used. In this paradigm, objects are represented as object-centered manner and recognition is performed by minimizing geometric variations.

Model-based approaches consist of alignment-based and invariant-based. Alignment based method tries to compensate the variability across views by finding optimal projection parameters as shown in Fig. 2.2 [19, 30, 52]. In this scheme, object recognition can be regarded as estimating camera model parameters from given 3D CAD model and point, line features extracted from 2D image. The projection parameters are estimated by the well-known hypothesis and test (verification) method. The idea is simply to hypothesize an object identity and its pose from all possible feature correspondences then render object model and verify the alignment iteratively. This method can provide exact object pose information but has several problems such as inefficient indexing for many objects and time consuming estimation process.

Figure 2.3. State-of-the art object representation: local visual feature-based method.

Invariant-based method tries to directly infer the presence of known object in the scene while minimizing the search using image invariants across views [65, 84, 85]. This approach stores a single object model (square dot in Fig. 2.2) regardless to the view variations. The invariants depend on the type of camera, such as affine camera and projective camera. The most well-known invariant in projective camera is 5 points invariant based on cross-ratio for planar shapes. This scheme can reduce the search space using efficient indexing, but it is only applicable to well-segmented 2D objects.

Since early 1990s, view-based or appearance-based recognition methods has been appeared for face and object recognition [66, 104]. These methods acquire all possible viewpoints and all possible illumination changes and store compressed data using PCA (principal component analysis) as shown in Fig. 2.2 right space (a set of images). They were also supported by psychology and neuro-physiological findings. Although this approach requires more storage than model-based approaches, it is relatively easy to learn and can recognize many objects. So it became the cornerstone of today's state-of-the art recognition system.

2.1.3   State-of-the art object recognition and related works

How to cope with the image variations caused by photometric, geometric distortions is one of the main issues in object recognition. It is generally accepted that the local invariant feature-based approach is very successful. This approach is generally composed of visual part detection, description, and classification. Fig. 2.3 shows the general visual feature generation flow by part detection and its description.

The first step of the local feature-based approach is the visual part detection. Schmid et al. [88] compared various interest point detectors and concluded that the scale-reflected Harris corner detector is most robust to image variations. Mikolajczyk and Schmid [62] also compared visual part extractor and found that Harris-

Laplacian based part detector is suitable for most applications. Recently, several visual descriptors are proposed [6, 33, 54, 63, 99]. Most approaches tried to encode local visual information such as spatial orientation or edge.

Based on these local visual features, several object recognition methods such as probabilistic voting method [87], constellation model-based approaches [22, 64] are introduced. Furthermore, there are more sophisticated classifiers such as SVM [113], Adaboost [122] that are basically based on the concept of nearest neighbors and their voting of local features. However, those approaches occasionally fail to recognize objects with few local features in highly cluttered backgrounds. Recently, Stein and Hebert [94] proposed a background invariant object recognition method by combining the object segmentation scheme. Since this method is based on prior figure-ground information, it cannot be used in general environment. Torralba et al. [102] introduced a completely different approach that exploits the background clutter information into object recognition. The background information is called scene context or exterior context in that method. The exterior context information is very useful for practical applications such as intelligent mobile robot system.

2.2   Objects in visual context

Objects do not exist alone but are interrelated to other visual components. We can categorize the visual context into three groups: spatial context, hierarchical context, and temporal context as shown in Fig. 2.4(a). Spatial context represents visual inter-action among spatial components such as part-part, object-object, object-region. Visual objects in a scene usually appear together and functionally related [5]. Recognition of an object provides expectations of other contextually related objects. This holds for visual parts in an object [58]. Solid arrows in Fig. 2.4(b) represent spatial interactions in object and part level. Spatial objects such as golf clubs, balls, and holes are strongly correlated. Spatial parts of eyes, a nose, and lips also interact together. We call such phenomena lateral or spatial interaction of visual components. The spatial interaction is another property of visual context which can also disambiguate object recognition during scene interpretation.

Hierarchical context represents visual interaction between different level of semantic concept such as part-object, object-scene. Scene context such as place information activates object recognition. Likewise, object recognition facilitates the scene recognition. This is bidirectional information exchange property of visual context [4]. The bidirectional interaction property exists in all perception layers in human brain. Block arrows in Fig. 2.4(b) show the bidirectional interaction between

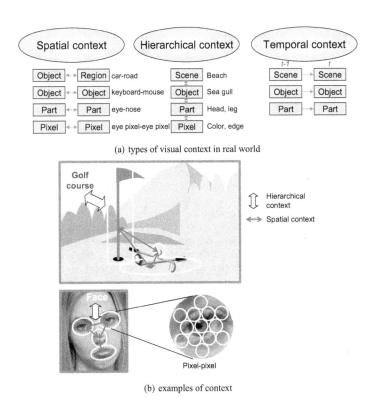

(a) types of visual context in real world

(b) examples of context

Figure 2.4. Categorization of visual context and examples.

scene-object layer (top) and object-part layer (bottom). If we see a golf course, we usually expect to see balls, holes, clubs, and etc. If we see golf balls, holes, and clubs, we convince that we are in a golf course. This example holds for object-part interaction such as face recognition. The bidirectional interaction of visual context is supported by the neuro-physiological facts proposed by Lee [45]. Lee found that bottom-up analysis and top-down synthesis exist in each visual processing area such as LGN, V1, V2, V4, IT, and higher cortical areas and processed concurrently.

Temporal context represents visual interaction between neighboring frames. In real world, a camera provides image sequences. Object information in previous frames provide strong prior of object existence in the next image.

2.3   Introduction of graphical model

As we discussed, the visual context provides relational information among visual components such as part, object, and scene. It is important to consider statistical variations in visual data due to several sources of noise in real environment. In general, modeling both the relation and uncertainty is very difficult problem. Fortunately, graphical models can do this job [32]. Graphical models combine both areas of graph theory (relation) and probability theory (uncertainty), and provide powerful, flexible framework for representing and manipulating global probability. We quote preface of [32] for clear concept of graphical model.

> Graphical models are a marriage between probability theory and graph theory. They provide a natural tool for dealing with two problems that occur throughout applied mathematics and engineering - uncertainty and complexity - and in particular they are playing an increasingly important role in the design and analysis of machine learning algorithms. Fundamental to the idea of a graphical model is the notion of modularity - a complex system is built by combining simpler parts. Probability theory provides the glue whereby the parts are combined, ensuring that the system as a whole is consistent, and providing ways to interface models to data. The graph theoretic side of graphical models provides both an intuitively appealing interface by which humans can model highly-interacting sets of variables as well as a data structure that lends itself naturally to the design of efficient general-purpose algorithms.

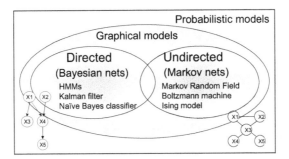

Figure 2.5. Scope of graphical model.

Graphical models are graphs where nodes represent random variables, and arcs represent probabilistic interaction between nodes. As shown in Fig. 2.5, there are directed graphical model and undirected graphical model according to the arc type. Directed graphical model is known as Bayesian net or causal model. Undirected graphical model is Markov net or Markov Random Field (MRF). In this section, we briefly introduce basics of graphical model for clear understanding of our work.

2.3.1   Directed graphical model

In a directed graphical model or Bayesian net, a direct arc from A to B represents that A causes B. The most important property of directed graphical model is the *conditional independence* relationship which simplifies the joint PDF. A node is independent of its ancestors given its parents. In Fig. 2.6, the joint PDF $P(A, B, C, D, E, F)$ is simplified if we use the conditional independence. It is just $P(A)P(B)P(C|A, B)$ $P(D|C)P(E|C)P(F|D, E)$. In general, Let $\{X\}$ denote the set of random variables for N nodes of the graph, $par\{X_i\}$ be the set of parents of node X_i, Then $P(X) = \prod_{i=1}^{N} P(X_i|par(X_i))$. We can see that the conditional independence relationships allow us to represent the joint PDF compactly. Most simpler forms of directed graphical models are Hidden Markov Model (HMM) and Kalman Filter.

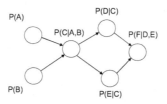

Figure 2.6. An example of directed graphical model.

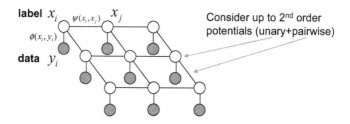

Figure 2.7. Graphical model of Markov Random Field (MRF) as undirected graphical model.

2.3.2 Undirected graphical model

For undirected graphical model, the basic subsets are cliques of the graph-subsets of nodes that are completely connected. For a given clique C, $\psi_C(x_C)$ denotes a general potential function-a function that assign a positive real number to each configuration x_C. Then joint PDF can be represented as $p(x) = 1/Z \prod_{C \in \mathbf{C}} \psi_C(x_c)$. $\mathbf{C}$ is the set of cliques and Z is normalization factor. A specific example of undirected graphical model is shown in Fig. 2.7. It is pairwise Markov Random Field (MRF) model. It considers up to 2nd order potentials (unary, pairwise). If $\{y\}$ is observation variables, $\{x\}$ is hidden variables, then the joint PDF can be written as $p(\{x\}, \{y\}) = 1/Z \prod_{(ij)} \psi_{ij}(x_i, x_j) \prod_i \phi_i(x_i, y_i)$. ψ is compatibility function between hidden variables and ϕ is evidence for measurement to hidden variable.

2.3.3 Inference

The main goal of inference is to estimate values of hidden variables given observation. In general, computing posterior using Bayes' rule is computationally intractable due to normalization constant Z except Gaussian case. Exact inference can be possible using variable elimination with NP-hard complexity. There are three kinds of approximate inferences such as sampling (Monte Carlo) methods, variational method, and belief propagation. You can see details in [32].

2.3.4 Learning

Learning includes either structure of model or parameter estimation. If a structure is known (fixed number of nodes and links) with full observability, then we can get closed form of learning. If partial observability (no label of observed data), we can run Expectation Maximization (EM) such as k-means clustering. If there is unknown structure, we have to find it using Bayesian inference with sampling. If we have no structure information and partial observability, we have to run structural EM such as BIC (Baysian Information Criterion) or MDL (Minimum Description Length) with EM. You can get much details in [15].

CHAPTER III
PIXEL CONTEXT: GENERALIZED ROBUST INVARIANT FEATURE

In this chapter, we present a new, biologically inspired perceptual feature to solve the scalability (selectivity) and generality (invariance) issue in object recognition. Based on the recent findings in neuronal and cognitive mechanisms in human visual systems, we develop a computationally efficient model. An effective form of a visual part detector combines a radial symmetry detector (convex part) with a corner-like structure detector (corner part). A general descriptor of pixel context encodes edge orientation, edge density, and hue information using a localized receptive field histogram. We compare the proposed perceptual feature (G-RIF: generalized robust invariant feature) with the state-of-the-art feature, SIFT, for feature-based object recognition. The evaluation results validate the robustness of the proposed perceptual feature in object recognition.

3.1 Selectivity and scalability of local feature

A successful object recognition system should have proper balance between selectivity and invariance. Selectivity means the system has to discriminate between different objects or parts. So it is related to the scalability issue. Invariance means that the same objects or parts have to be invariant to unknown photometric and geometric variations. This is related to the generality issue. It is generally accepted that the local invariant feature-based approach is very successful in this aspect. This approach is generally composed of visual part detection, description, and classification as introduced in Chapter 2.

3.2 Mechanisms of Receptive Field

3.2.1 Simple cells/complex cells in V1

Simple cells and complex cells exist in the primary visual cortex (V1), which detects low level visual features. It is well known that the response of simple cells in V1 can be modeled by a set of Gabor filters as Eq. 3.1 [90]:

$$G(x, y, \theta, \varphi) = e^{-\frac{(x'^2 + \gamma y'^2)}{2\sigma^2}} cos(2\pi \frac{x'}{\lambda} + \varphi) \qquad (3.1)$$

where $x' = xcos\theta + ysin\theta$ and $y' = -xsin\theta + ycos\theta$. According to recent neurophysiological findings [82], the range of an effective spatial phase parameter is $0 \leq \varphi \leq \pi/2$ due to symmetry. An important finding is that the distribution of spatial phase is bimodal, with cells clustering near 0 phase (even symmetry) and $\pi/2$ phase (odd symmetry).

Complex cell response at any point and orientation combines the simple cell responses. There are two kinds of models: weighted linear summation and MAX operation [90]. However, the MAX operation model has the most support since neurons performing a MAX operation have been found in the visual cortex [25]. The role of the simple cell can be regarded as a tuning process, that is, extracting all responses by changing Gabor parameters. The role of a complex cell can be regarded as a MAX operation from the tuning responses by selecting maximal responses. The former gives selectivity to the object structure and the latter gives invariance to geometric variations such as location and scale changes.

3.2.2 Receptive field in V4

Through the simple and the complex cells, orientation response maps are generated and fine orientation adaptation occurs on the receptive field within the attended convex part in V4 [9]. The computational method of orientation adaptation phenomenon is steering filtering [11]. Adapted orientation is calculated by the maximum response spanned by basis responses ($tan^{-1}(I_y/I_x)$). There are also color blobs in a hyper column where the opponent color information is stored. Hue is invariant to affine illumination changes and highlights. Orientation information and color information is combined in V4 [118].

How does the human visual system (HVS) encode the receptive field responses within the attended convex part? Few facts are known on this point, but it is certain that larger receptive fields are used, representing a broader orientation band [39].

3.3   Visual Part Detection

What is a good object representation scheme comprising both selectivity and invariance? The global description with the perfect segmentation may show very good selectivity. However, this representation does poorly with respect to invariance since a perfect segmentation is impossible, and further, is sensitive to visual variation in light, view angle, and occlusion. Objects representation as the sum of sub-windows or visual parts may be a plausible solution and supported by the recognition by component (RBC) theory [8]. The main issues of this approach are how to select the location, shape, and size of a sub-window and what information to be encoded for both selectivity and invariance.

In this section, we present a visual part detection method by applying the tune-MAX [90] to the approximated Gabor filter. Serre and Riesenhuber modeled the Gabor filter using a lot of filter banks while changing scale and orientation. Furthermore, they fixed the phase to 0 which is another limitation. As described in Sect. 3.2.1, the distribution of spatial phase in receptive field is bimodal, 0 (even symmetry) and $\pi/2$ (odd symmetry). So, we can approximate the Gabor function by two bases generated by the Gaussian derivatives shown in Fig. 3.1(a) 3.1(b). The 1st and 2nd derivatives of Gaussian which approximate odd and even symmetry, respond to edge structures and convex (or bar) structures respectively.

The location and size invariance is acquired by the MAX operation of various tuning responses from the 1st, 2nd derivatives of Gaussian. Fig. 3.2 shows the complex cell responses using the approximated filters. The arrows represent the tuning process and the dot and circle represents the selected feature position and region size by the MAX operation.

(1) Location tuning using the 1st derivative of Gaussian:

- Select maximal response in all orientations within a 3×3 complex cell (pixel).

- Suitable method: Harris corner or KLT corner extraction (both eigenvalues are large)

(2) Location tuning using the 2nd derivative of Gaussian:

- Select maximal response in all orientations within 3×3 complex cell.

- Suitable method: Laplacian or DoG gravity center (radial symmetry point)

(3) Scale tuning using the convexity:

- Select maximal response in directional scale-space [75].

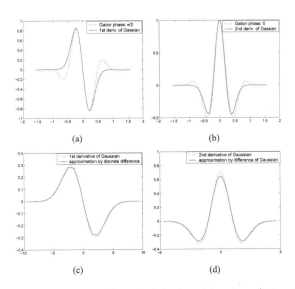

Figure 3.1. Gabor filters are approximate by derivatives of Gaussian: $\pi/2$ phase Gabor by 1st deriv. of Gaussian (a) and 0 phase Gabor by 2nd deriv. of Gaussian (b). The 1st and 2nd deriv. of Gaussian are further approximated by discrete difference of Gaussian (c) and by difference of two Gaussian kernels, respectively for computational efficiency.

Figure 3.2. (Left two) Interest points are localized spatially by tune (arrow)-Max (dot) of 1st and 2nd derivative of Gaussian respectively. (Right two) Region sizes around interest points are selected by tune-Max of convexity in scale-space.

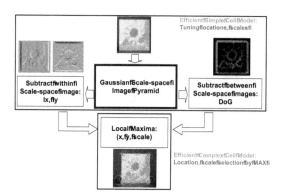

Figure 3.3. Computationally efficient perceptual part detection scheme

- For computational efficiency, a convexity measure such as DoG is suitable. This is related to the properties of V4 receptive field where convex part is used to represent visual information [79]

For the efficient computation of the tune-Max, we utilize three approximation schemes: the scale-space based image pyramid [54], discrete approximation of the 1st derivative of Gaussian by subtracting neighboring pixels in a Gaussian smoothed image, and the approximation of the 2nd derivative of Gaussian (Laplacian) by difference of Gaussian (DoG). As shown in Fig. 3.1(c) 3.1(d), the kernel approximations for the Gabor bases are almost identical to the true kernel function. Fig. 3.3 shows the structure of our part detection method which computes the 1st and 2nd derivatives of Gaussian by subtracting neighboring pixels in a scale-space image and by subtracting between scale-space images, respectively. We calculate local max during scale selection. Fig. 3.4 shows a sample result of the proposed perceptual part detector. Note that the proposed method extracts complementary visual parts. We can get corner-like parts through the left path, and radial symmetry parts through the right path in Fig. 3.3 (See Eq. 3.2). As shown in Fig. 3.5, the proposed object decomposition method is also supported by cognitive properties that HVS attends to gravity centers and high curvature points [81] and objects are deconstructed into perceptual parts that are convex [42, 55].

$$\mathbf{x} = \max_{\mathbf{x} \in W}\{DoG(\mathbf{x}, \sigma) or HM(\mathbf{x}, \sigma)\}, \sigma = \max_{\sigma}\{DoG(\mathbf{x}, \sigma)\} \qquad (3.2)$$

where $DoG(\mathbf{x}, \sigma) = |I(\mathbf{x}) * G(\sigma_{n-1}) - I(\mathbf{x}) * G(\sigma_n)|$ and $HM(\mathbf{x}, \sigma) = det(\mathbf{C}) - \alpha trace^2(\mathbf{C})$. $\mathbf{C}$ is defined as Eq. 3.3.

$$\mathbf{C}(\mathbf{x}, \sigma) = \sigma^2 \cdot G(\mathbf{x}, 3\sigma)) \cdot \begin{bmatrix} I_x^{\ 2}(\mathbf{x}, \sigma) & I_x I_y(\mathbf{x}, \sigma) \\ I_x I_y(\mathbf{x}, \sigma) & I_y^{\ 2}(\mathbf{x}, \sigma) \end{bmatrix} \qquad (3.3)$$

where $I_x(\mathbf{x}, \sigma) = \{S([x+1, y], \sigma) - S([x-1, y], \sigma)\}/2$, $I_y(\mathbf{x}, \sigma) = \{S([x, y+1], \sigma) - S([x, y-1], \sigma)\}/2$, $S(\mathbf{x}, \sigma) = I(\mathbf{x} * G(\sigma))$

3.4   Perceptual Part Descriptor

As we discussed in Sec. 3.2.2, we can mimic the role of receptive field V4 to represent visual parts. In V4, edge density map, orientation field, and hue field coexist in the attended convex part. These independent feature maps are detected from V1 (in particular, edge orientation and edge density is extracted using the approximated Gabor with $\pi/2$ phase).

Now the question becomes: How to encode the independent feature maps? We utilize the fact that larger receptive fields are used with a broader orientation band [39] and independent feature maps are combined to make more informative features [118]. Fig. 3.6(a) shows a possible pattern of receptive field in V4. The density of the black circle depicts the level of attention of the HVS in which 86% of fixation occurs around the center receptive field [81]. Each black circle stores the visual distributions of edge density, orientation field, and hue field of pixels around the circle. Fig. 3.6(b) shows how it works. Each receptive field stores them in the form of a histogram which shows good balance between selectivity and invariance by controlling the bin size, and is partially supported by the computational model of multidimensional receptive field histograms [86]. We can control the resolution of each receptive field such as the number of edge orientation bins, hue orientation bins except edge density which is scalar. Each localized sensor gathers information about edge density, edge orientation, hue color of receptive field. The histogram of an individual sensor is generated by simply counting the corresponding bins according to feature values weighted by the attention strength and sensitivity of sensor (modeling of pixel context). Scalar edge density is generated from edge magnitudes. This process is linear to the number of pixels within a convex part. Each pixel in a receptive field affects to neighboring visual sensors. After all the receptive field

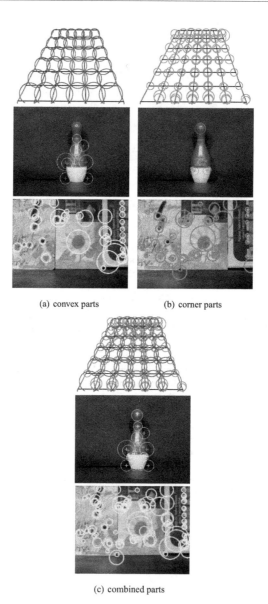

(a) convex parts (b) corner parts

(c) combined parts

Figure 3.4. The proposed part detector can extract convex, corner part simultaneously. (1st row) geometric structure, (2nd row) man-made object, (3rd row) natural scene

 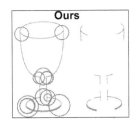

Figure 3.5. Object decomposition comparison between SIFT [54] and our model.

Table 3.1. Comparison of dominant orientation detection methods

Method	Eig. vector	Steerable filter	Ori. hist.	Radon tr.
# of correct/# of tot.	12/20	16/20	9/20	10/20
Matching rate [%]	60 %	**80**%	45 %	50 %

histograms are generated, we normalize each histogram vector. After we multiply these histograms with component weights ($\alpha + \beta + \gamma = 1$), we integrate three kinds of features as Fig. 3.6(b) right column. Finally, we renormalize the feature (dim.: 21*(4+1+4)=189) so that the feature's energy equals to 1.

It is essential to align the receptive field patterns (shown as black circles in Fig. 3.6(a)) to the dominant orientation of an attended part if there is image rotation. We compared four dominant orientation detection methods: Eigenvector [56], weighted steerable filter [11], maximum of orientation histogram [54], and radon transform [31]. The eigenvector method calculates the dominant orientation from an eigenvector of a Hessian matrix. The weighted steerable filter calculates the dominant orientation using weighted gradient components (x-direction and y-direction). The orientation histogram method uses max orientation bins generated from pixel orientations. Variance projection in Radon transform can provide dominant orientation information. We found that the weighted steerable filter method showed the best matching rate in rotated image pairs, as shown in Table 3.1.

Each receptive field encodes histogram-based orientation distribution in $[0, \pi]$, hue distribution in $[0, 2\pi]$ and scalar value of edge density in $[0, 1]$. The coordinates of the receptive field are aligned to the dominant orientation of the convex visual part, which is calculated efficiently by steering filtering [11]. This encoding scheme can control the level of local context information-aperture size and feature level as

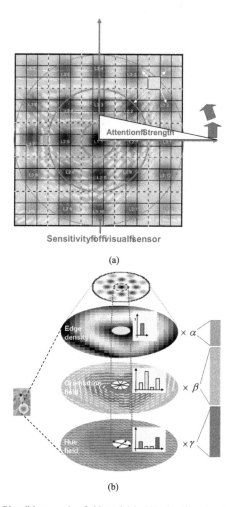

(a)

(b)

Figure 3.6. (a) Plausible receptive field model in V4, (b) Three kinds of localized histograms are integrated to describe local region.

Table 3.2. Level of local context.

Context	Contents of information
Aperture size	L1:1 L2:1,L2:2,L2:3,L2:4,L2:5,L2:6,L2:7,L2:8 L3:1,L3:2,L3:3,L3:4,L3:5,L3:6,L3:7,L3:8, L3:9,L3:10,L3:11,L3:12 L4:1,L4:2,L4:3,L4:4
Type of feature	(1) Edge orientation $[0, \pi]$, quan. level:4 (2) Edge density $[0, 1]$ (3) Hue information $[0, 2\pi]$, quan. level:4 etc: (1)+(2), (1)+(3),(2)+(3), (1)+(2)+(3)

in Table 3.2. The proposed encoding scheme is a general form of the contextual feature representations of [6, 54, 99]. We can select a suitable level of context depending on applications. In general, as the aperture size is larger, the recall is lower and the precision is higher. Figure 3.7 shows an example of visual part matching using the proposed part detector and general context descriptor (G-RIF: generalized robust invariant feature). Most visual parts correspond to one another by a simple Euclidean distance measure. More specific details of implementation and invariant properties can be found in [35].

3.5   Evaluation of G-RIF

We dubbed the proposed perceptual feature (perceptual part detector with generalized descriptor of pixel information) as the Generalized Robust Invariant Feature (G-RIF). In this section, we evaluate G-RIF in terms of object recognition. We adopt a new feature comparison measure in terms of object labeling. We use the accuracy of detection rate which is widely used in classification or labeling problems. Although there are several suitable open object databases such as COIL-100 and Caltech DB, we evaluate the proposed method using our own database because our research goal is to measure the properties of features in terms of scale change, view angle change, planar rotation, illumination intensity change, and occlusion. The total number of objects is 104: related test images are shown in Fig. 3.8 and the

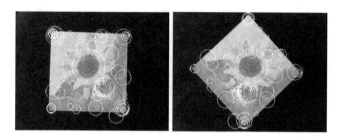

Figure 3.7. The matching results on the rotated objects using the perceptual part detector and local contextual descriptor [G-RIF: aperture size L3 and feature type (1)+(2)].

104 object models are shown in Fig. 3.13. These DB and test images are acquired using a SONY F717 digital camera and resized to 320×240.

We compare the performance of the proposed G-RIF with SIFT, the state-of-the art feature [54]. We evaluate the features using the nearest neighbor classifier with direct voting (NNC-voting) which is used commonly in local feature-based object recognition approaches. NNC-based voting is a very similar concept to the winner-take-all (WTA). We use the binary program offered by Lowe [54] for the accurate comparison.

Fig. 3.9 shows the performance of the proposed perceptual part detectors. We use the same descriptor (edge orientation only) with the same Euclidean distance threshold (0.2) used in NNC-based voting. The proposed perceptual part detector outperforms single part detectors in most test sets. The maximal recognition rate is higher than the part detector of SIFT (radial symmetry part) by 15%. Fig. 3.10 shows the performance of the proposed part descriptor with the SIFT descriptor. We used the same radial symmetry part detector to show the power of descriptors only. The full contextual descriptor (edge orientation + edge density + hue field) shows the best performance except the illumination intensity change set. In this case, the performance is fair compared to the other contextual descriptors. Under severe illumination intensity change or different light sources, it is reasonable to use the contextual descriptor of edge orientation with edge density.

Fig. 3.11 summarizes the performance of the SIFT, perceptual part detector with edge orientation descriptor, and the G-RIF (both parts with general descriptor). The G-RIF always outperforms the SIFT in all test sets. This good performance is originates from the effective use of image structures (radial symmetry point with a high

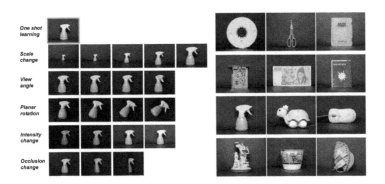

Figure 3.8. We use frontal 104 views for DB (right) generation and 20 test set (5 scales, 4 view angles, 4 rotation, 4 intensity change, 3 occlusions) per object (left).

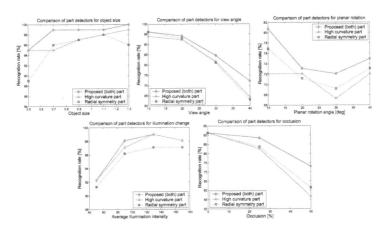

Figure 3.9. Evaluation of part detectors (radial symmetry part, high curvature part, perceptual part): The proposed part detector shows the best performance for all test set.

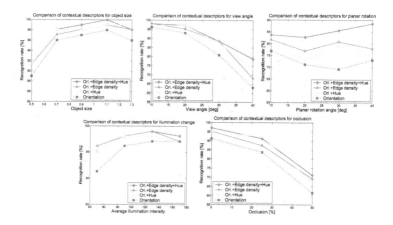

Figure 3.10. Evaluation of part descriptors (ori. only, ori+hue, ori+edge, ori+hue+edge): The full descriptor shows almost best performance.

curvature point) of the proposed visual part detector and effective spatial coding of multiple features in a unified way. The dimension of G-RIF is 189 (21*4+21*4+21) and that of SIFT is 128 (16*8). The average extraction time of G-RIF is 0.15sec and that of SIFT is 0.11sec in a 320*240 image under AMD 2400+. This difference is due to the number of visual parts. The G-RIF extracts twice number of parts than the SIFT does.

We also compared G-RIF and SIFT on COIL-100 DB [70, 71]. Since the DB is composed of multiple views with an interval of 5°, we use the frontal view (0°) as the model image and tested five different views up to 45°. We use the nearest neighbor classifier (NNC) based simple voting as the classifier. Fig. 3.12 shows that G-RIF has upgraded performance (by 20%) at the 45°viewing angle.

3.6 Summary

In this chapter, we introduced a Generalized-Robust Invariant Feature which shows good performance in terms of selectivity (recognition accuracy) and invariance (to various test images). First, we detect perceptually meaningful visual parts de-

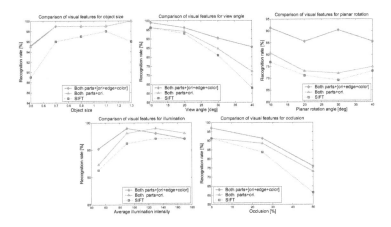

Figure 3.11. Summary of visual features (SIFT, perceptual part + edge orientation, G-RIF): G-RIF shows the best performance. Both parts+ori and SIFT follow G-RIF.

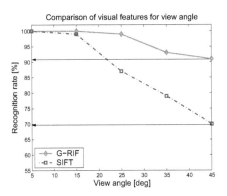

Figure 3.12. G-RIF vs. SIFT for COIL-100 DB. G-RIF is more robust than SIFT for viewing angle changes by 20%. We used the same voting-based classifier.

rived from the properties of the visual receptive field of V1. Applying the Tune-MAX scheme to two basis Gabor kernels can extract complementary visual parts (Fig. 3.4). Second, we also proposed a generalized contextual encoding scheme based on the properties of receptive field V4 and attention of the HVS. The information of edge field and hue field is characterized by the localized histogram weighted according to the attentional strength. It is a generalized form of SIFT descriptor, shape context, minutia descriptor. The performance of the G-RIF compared with the state-of-the art feature shows the recognition power using the NNC-based simple voting. Effective utilization of image structures and pixel information give good performance in feature-based object recognition.

This chapter shows object labeling method by utilizing only the pixel context for segmented or no cluttered background. Since this is the simplest situation of object recognition, it shows good labeling performance with pixel context (G-RIF). Chapter 4 presents how to utilize part context for object labeling in cluttered background.

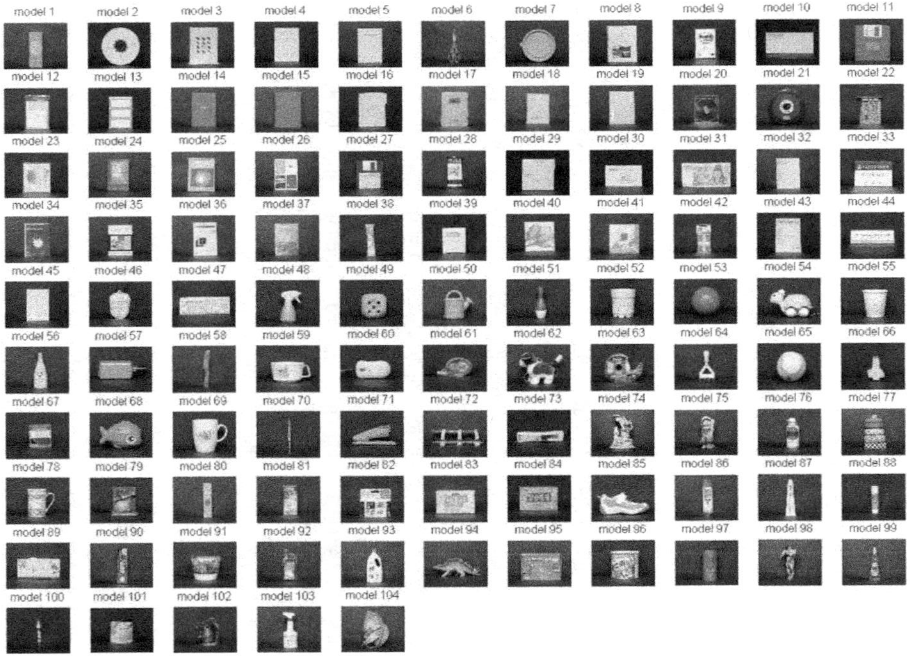

Figure 3.13. Composition of KAIST-104 object database.

CHAPTER IV

PART CONTEXT: BACKGROUND ROBUST OBJECT LABELING

In this chapter, we introduce part-part context which is one of spatial context and propose a mathematical method which shows robustness to background clutter. The conventional simple voting of local feature shows very poor performance under background clutter since it assumes independent feature. To reduce the effect of clutter, first we utilize the G-RIF for object decomposition and description. Then, we incorporate part-part context in the form of weight aggregation between the features and the neighboring features using similarity and proximity. Through the recursive weight aggregation process, features belonging to the same objects get stronger weights, and features belonging to clutter get weaker weights. This is similar to belief propagation. Then, we vote the weight-aggregated features to get the object labels. We validate the enhancement of the proposed method by using an open database and a real test set. This proposed method is inspired by the function of the human visual system, called figure-ground discrimination. We use the proximity and similarity between features to support each other. The contextual feature descriptor and contextual voting method, which use contextual information, enhance the recognition performance enormously in severely cluttered environments.

4.1 Background clutter problem

How to cope with image variations caused by photometric and geometric distortions is one of the main issues in object recognition. It is generally accepted that the local invariant feature-based approach is very successful in this regard. This approach is generally composed of visual part detection, description and classification.

The first step in the local feature-based approach is visual part detection. Lindeberg [51] proposed a pioneering method on blob like image structure detection in scale-space. Shokoufandeh [92] extended this feature to wavelet domain. Schmid et al [88] compared various interest point detectors and concluded that the scale-reflected Harris corner detector is most robust to image variations. Mikolajczyk and Schmid [62] also compared visual part extractors and found that the Harris-Laplacian based part detector is suitable for most applications. Recently, several visual descriptors have been proposed [33, 54, 63, 99]. Most approaches try to encode local visual information such as spatial orientation or edge.

Based on these local visual features, several object recognition methods, such as the probabilistic voting method [87] and constellation model-based approaches [22, 64], have been introduced.

However, those approaches occasionally fail to recognize objects with few local features in highly cluttered backgrounds. What happens if we take conventional object recognition systems out of controlled dark rooms to real, cluttered environments? Even successful object recognition systems in dark rooms often fail to recognize objects when distracted by clutter in real environments. How can we reduce or minimize the influences of background clutter on object recognition? This is the main problem to solve in this chapter.

Several researchers have tried to solve the background clutter problem. Stein and Hebert [94] proposed a background invariant object recognition method by combining the object segmentation scheme with SIFT. Since this method is based on the prior figure/ground information using manual segmentation or stereo matching, this method cannot be used in the general environment. Recently, object-specific, figure-ground segmentation methodologies by combining bottom-up and top-down cues have been proposed [41, 47, 120]. Although these approaches show somewhat clear figure-ground segmentation results using object-specific information, there is no concern about how to reduce the influence of background clutter for successful object recognition in a non-segmentation-based approach. Only Lowe [54] tried to discard background clutter by using a simple distance ratio, which in practice is useful. Torralba et al. [102] introduced a completely different approach that exploits the background clutter information into object recognition. The background information is called place context or exterior context in that method. The exterior context information is very useful for practical applications such as intelligent mobile robotics systems.

Our research interest is how to efficiently extract an object's interior contextual information to enhance object recognition performance in strongly cluttered environments. We concentrate on the properties of a human visual receptive field and propose a computationally efficient local context coding method. We also propose an object recognition method based on the human visual system's figure-ground discrimination mechanism, which is accomplished by grouping interior contextual information. Fig. 4.1 summarizes the clutter reduction strategy proposed in this

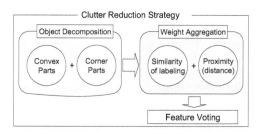

Figure 4.1. Clutter reduction strategy.

chapter. The first key idea is to decompose an object into convex parts and corner parts in order to include fewer background pixels. The second key idea is to aggregate feature weights according to similarity and proximity. Parts of an object tend to share the same attributes and aggregate together. We can discard background clutter by voting the aggregated weights.

4.2 Object decomposition and description

Appearance- or view-based object recognition methods were proposed in the early 1990s for face recognition and currently have become popular in the object recognition society. However, global appearance representation using a perfect support window as shown in Fig. 4.2(a) cannot recognize objects in cluttered environments because of imperfect figure-ground segmentation. A bypassing method is to approximate an object as the sum of its sub-windows. The main issue in this approach is how to select the location, shape and size of a sub-window as shown in Fig. 4.2(b). For the successful object recognition, visual parts have to satisfy the following requirements. First, background information should be excluded as much as possible. Second, they must be perceptually meaningful. Finally, they should be robust to the effects of photometric and geometric distortions. Mahamud [57] suggested a greedy search method to find visual parts that satisfy the above requirements. Image structure-based part detectors are more efficient than regular grid-based part [54, 88]. Although these methods work well, they do not exploit the full structural information of interesting objects.

We utilize the perceptual feature, G-RIF explained and evaluated in Chapter 3. It decomposes an object into convex parts and corner-like interior parts to fully

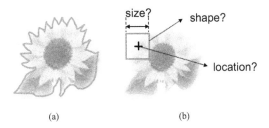

(a)   (b)

Figure 4.2. (a) Perfect global object description. (b) Issues in local part-based object description.

utilize the structural information. The proposed object decomposition method is supported by biological facts [42]. Convex parts include little background area and are geometrically stable to view point changes. There are also corner-like interior parts between the convex parts. Corner-like parts provide structural glues among the convex parts within the object region. We detect the convex parts using Laplacian and the corner parts using the well-known Harris corner in scale space [35]. After object decomposition, we have to characterize the detected parts to represent local appearance. Pixels in a part have gradient magnitude, orientation, and hue. We encode these distributions by localized histograms. We build orientation, density, and hue histograms at each small blob. The localized histograms are integrated into a feature vector. The G-RIFs are used as inputs to the voting stage.

4.3   Part-part context based object recognition

A conventional classifier based on local informative features is direct voting of nearest neighbors as in Fig. 4.4(a) [87]. Although there are more sophisticated classifiers, such as SVM [113], Adaboost [122], Bayesian decision theory [64], and strong spatial constraint-based indexing [91], those methods are basically based on the concept of nearest neighbors and their voting.

Basically, we consider two properties of part relation: the same labeling and proximity as shown in Fig. 4.3. Parts belonging to an object share the same object labels. Furthermore, those parts are spatially very close. Proper weights are assigned to those parts according to the probability of the same labeling and proximity. Our research interest is how to utilize the part relation in the nearest-neighbor

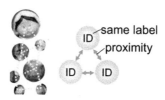

Figure 4.3. Similarity and proximity of part-part context.

voting scheme (top-down verification is beyond the scope). Contextually supported parts get stronger weights with a certain label. We can estimate object labels by voting the weights. This is the reason why this is called weight-aggregated voting.

The proposed object recognition method, named neighboring context-based voting, can be modeled as follows:

$$L = \arg\max_{l} P(l|\mathbf{X}) \approx \arg\max_{l} \sum_{i=1}^{N_\mathbf{x}} P(\mathbf{x}_i|l) \qquad (4.1)$$

where local feature $\mathbf{x}_i$ belongs to input feature set $\mathbf{X}$, l is an object label and $N_\mathbf{x}$ is the number of input local features. The posterior $P(l|\mathbf{X})$ is approximated by the Bayes rule and uniform prior of label. We use the following binary probability model to design $P(\mathbf{x}_i|l)$:

$$P(\mathbf{x}_i|l) = \begin{cases} 1 & i \in l, S_M(i,l) \geq \epsilon \\ 0 & otherwise \end{cases} \qquad (4.2)$$

where $S_M(i,l)$ is called a feature-label map which represents the strength of the match between an input feature and its label based on neighboring context information. It is calculated as follows:

$$S_M(i,l) = \sum_{j \in N(i)} w(i,j)c(j,l) \qquad (4.3)$$

where $N(i)$ is the support region (or neighbors) of feature i. As shown in Fig. 4.5, neighboring context information is aggregated by summing over the neighborhood using the above equation. $w(i,j)$ represents the support (contextual) weight at site j in the support region. The support from the neighborhood is valid when the neighboring visual parts have the same label (or object index) as the site of interest. This

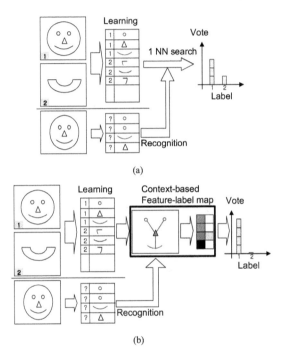

(a)

(b)

Figure 4.4. (a) Conventional classifier: direct voting of nearest neighbor. (b) Novel classifier: neighboring context-based voting.

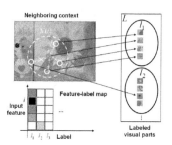

Figure 4.5. oncept of feature-label map (strength of match) generation using neighboring context information.

is the same concept as Gestalt properties, i.e., proximity and similarity. $c(j, l)$ is the goodness of match pair (j, l).

The support weight will be proportional to the probability, $p(O_j^l | O_i^l)$ where O_i^l represents the event that the label of a site i is l:

$$w(i, j) = k \cdot p(O_j^l | O_i^l) \qquad (4.4)$$

When we consider the fixed label l, it can be rewritten as:

$$w(i, j) = k \cdot exp\left\{ -\frac{dist(\mathbf{x}_i, \mathbf{x}_j)^2}{\sigma^2} \right\} \cdot p(l\mathbf{x}_i = l, l\mathbf{x}_j = l) \qquad (4.5)$$

where the first term represents the proximity between two parts and the second term represents the similarity of the same label. $dist(\mathbf{x}_i, \mathbf{x}_j)$ denotes the distance between part i and j. $l\mathbf{x}_i$ represents the label of feature $\mathbf{x}_i$ and k is normalization factor. The probability of both features coming from the same object l is defined as:

$$p(l\mathbf{x}_i = l, l\mathbf{x}_j = l) = \frac{S(l\mathbf{x}_i = l)S(l\mathbf{x}_j = l)}{\sum_{m=1}^{N_L} \sum_{n=1}^{N_L} S(l\mathbf{x}_i = n)S(l\mathbf{x}_j = m)} \qquad (4.6)$$

where $S(l\mathbf{x}_i = l)$ means the similarity that a local feature is mapped to a label l. This is equivalent to the χ^2 kernel value between input feature $\mathbf{x}_i$ and its nearest neighbor x_{1NN}^l, whose label is l:

$$S(l\mathbf{x}_i = l) = K_{\chi^2}(\mathbf{x}_i, x_{1NN}^l) \qquad (4.7)$$

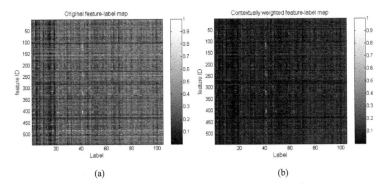

(a)   (b)

Figure 4.6. (a) Initial feature-label map obtained from similarity kernel. (b) Contextually weighted feature-label map using the proposed method.

The χ^2 similarity kernel is defined as [113]:

$$K_{\chi^2}(\mathbf{x}, (y)) = exp\{-\rho\chi^2(\mathbf{x}, \mathbf{y})\} \tag{4.8}$$

where ρ is set to 3~4 normally. Goodness of match, $c(j, l)$ is also defined as $S(l\mathbf{x}_j = l)$.

Although the feature-label map $S_M(i, l)$ is calculated using the above similarity kernel, it can be formulated in recursive form by simply reusing the feature label map in the next iteration, rather than the similarity kernel.

Fig. 4.6(a) shows the feature label map calculated from the similarity kernel and Fig. 4.6(b) shows the feature label map after applying the proposed method: contextually weighted feature-label map after two recursions. Fig. 4.7 represents a close up view of Fig. 4.6 and displays several feature locations. Note the power of neighboring contextual information, which enhances the strength of figural features (here, f.id: 527, 536) and suppresses the background features (here, f.id: 525).

The baseline distance measure of voting is Euclidean feature distance. In this case, $logP(l|(x_i) = 1$ if the label of nearest (x_i) is l. Fig. 4.8(a), 4.8(b) show the feature labeling and voting results respectively. As you can see, many background clutters are labeled incorrectly. The next improved distance measure for clutter rejection is distance ratio obtained by comparing the distance of the nearest to that of the second closest neighbor [54]. Fig. 4.8(c), 4.8(d) show the feature labeling and voting results respectively. The incorrectly labeled features are removed partially.

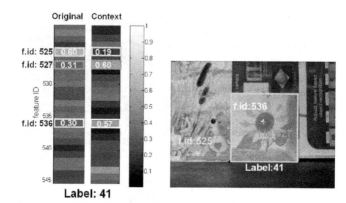

Figure 4.7. Close-up view of Fig. 4.6 around Label 41 and corresponding feature locations.

Finally, we apply the weight aggregation-based distance measure to the same test image. Fig. 4.8(e), 4.8(f) show the feature labeling and voting results respectively. Note that all the object-related features are labeled correctly. In this test, we tune the related distance thresholds so that the number of correctly labeled feature is almost same.

4.4   Evaluation of contextual voting

Although there are several open object databases, such as Amsterdam DB, PAS-CAL DB, ETH-80 DB, and Caltech DB, we evaluate the proposed method using COIL-100 DB for feature comparison. We also use our DB, CMU DB (available at http://www.cs.cmu.edu/~stein/BSIFT) which contain background clutter, because our research goal is to recognize general objects in severely cluttered environments. Fig. 4.9 (top) shows sample objects in our DB, composed of 21 textureless 2D objects, 26 textured 2D objects, 26 textureless 3D objects and 31 textured 3D objects. We stored each model's image in the canonical viewpoint. Because our research issue is how the contextual information enhances recognition, fixing a 3D viewpoint is reasonable. To effectively validate the power of the context-based object recog-

Table 4.1. Comparison of labeling performance for KAIST-104 DB.

Feature	SIFT	G-RIF	G-RIF	G-RIF
Distance	Nearest	Nearest	Ratio	Weight
# correct/ # test	62/104	68/104	70/104	74/104
Success rate [%]	59.61%	65.38%	67.31%	71.15%

nition method, we made test images as in Fig. 4.9 (bottom). 104 objects are placed on a highly cluttered background. These DB and test images are acquired using a SONY F717 digital camera and are resized to 320×240.

First, we compare the performance of the proposed G-RIF with SIFT, which is a state-of-the art descriptor [54,63]. Typically, visual part detectors are evaluated only in terms of their repeatability. However, the evaluation should be in the context of a task, such as accuracy of object labeling in this chapter. We evaluate the features using the direct voting-based classifier which is used commonly as in Fig. 4.4(a). We use the binary program offered by Lowe [54]. The accuracy of detection rate is used as a comparison measure which is widely used in classification or labeling problems. Fig. 4.10 shows the evaluation results using our DB by changing the similarity threshold of direct voting. G-RIF shows better performance than SIFT and reveals the complementary properties of G-RIF: right path (gravity center point) and G-RIF: left path (high curvature point).

The 2nd and 3rd columns in Table 4.1 show the overall recognition results using the optimal classifier threshold (0.8). The recognition rate using G-RIF is higher than the SIFT feature by 5.77%. This good performance originates from the complementary properties of the proposed visual part detector and the effective spatial coding of multiple features in a unified way. In this test, we set the aperture level up to L3 and feature level by (1)+(2). Next, we compared our weight aggregation-based voting with the two baseline methods, nearest-neighbor and distance ratio with the same G-RIF. The neighborhood size is set to 2 times the part scale. Overall test results are shown in Table 4.1 (3rd, 4th, 5th column). The weight aggregation-based classifier shows the best recognition performance among other baseline methods. This good performance originated from the effective use of neighboring context. Fig. 4.11 shows several successful examples using G-RIF that are incorrect using the SIFT feature, although our feature dimension is 105 (84 for edge orientation + 21 for edge density) and SIFT's is 128.

Next, we compared our context-based voting (Fig. 4.4(b)) to the conventional nearest-neighbor based direct voting (Fig. 4.4(a)). We used the same visual feature (G-RIF) proposed in Chapter 3 and our DB. The neighborhood size is set to two times the part scale. Overall test results are shown in Table 4.1 (3rd and 4th columns). The contextual voting-based classifier shows better recognition performance than direct voting by 5.77%. This good performance originates from the effective use of the neighboring context. The contextual influence was explained in Fig. 4.7. Fig. 4.12 shows three kinds of recognition examples with both methods. Fig. 4.12(a) is a normal case. Both methods succeeded for textured objects. Fig. 4.12(b) show ambiguous recognition results by direct voting but stable recognition results by contextual voting. Finally, Fig. 4.12(c) shows the failure case by direct voting, but success with contextual voting. Fig. 4.13 shows other results of Fig. 4.12(c)'s case. Due to many ambiguous features on the cluttered background, the direct voting method labels the wrong object. However, the context-based voting method can reduce the effect of those cluttered features by neighboring contextual information. Finally, we evaluated contextual and direct voting using the CMU database. This database is composed of 110 separate objects and 25 background images. We generated test images by changing objects' sizes on different backgrounds. This is equivalent to changing the background size. Fig. 4.14 shows the evaluation results. Contextual voting shows an equal or better recognition rate than direct voting. Note that the neighboring contextual information has a more important role for object labeling in a severely cluttered background than in a less cluttered background. Finally,

4.5 Summary

In this chapter, we propose a simple and powerful object labeling method based on part-part context. It is modeled by weight aggregation. Such spatial part context suppresses the strength of background features and enhances the strength of figural features in the feature-label map. The proposed method shows better labeling performance than other methods in a severe environment.

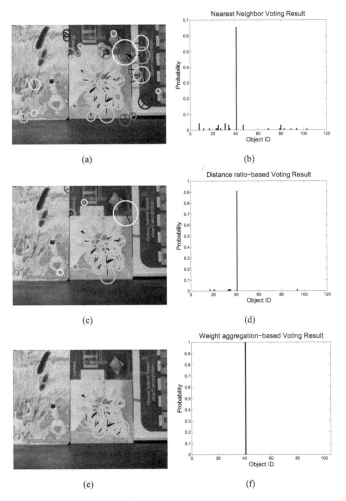

Figure 4.8. Feature labeling and voting using nearest neighbor (a)(b), distance ratio (c)(d), and the proposed weight aggregation (e)(f).

Figure 4.9. (Top) examples of KAIST-104 general object models, (bottom) corresponding examples of KAIST-104 test objects on a highly cluttered background.

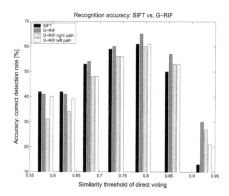

Figure 4.10. Comparison of NNC classification accuracy between G-RIF (left, right, both paths) and SIFT.

Figure 4.11. Correct detection using G-RIF (failure cases using SIFT).

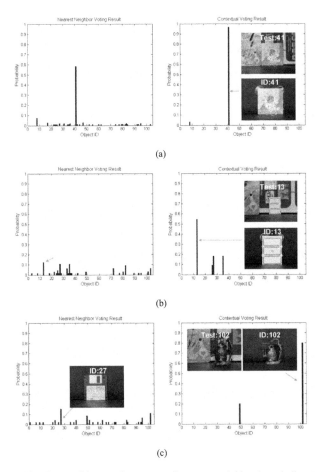

Figure 4.12. Recognition performance of nearest neighbor-based direct voting method (left column) and contextual voting method (right column): (a) Normal case, (b) ambiguous case and (c) failure using direct voting but success using ours.

Figure 4.13. Correct detection using contextual voting (failure cases using direct voting).

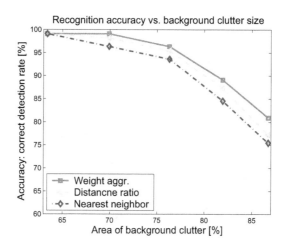

Figure 4.14. As the area of background clutter increases, the contextual voting shows relatively higher detection accuracy than the direct voting.

CHAPTER V

PART-OBJECT CONTEXT: SCALABLE 3D OB-JECT IDENTIFICATION AND LOCALIZATION

In Chapter 3, we proposed robust local feature (G-RIF) where pixel context is encoded by a localized histogram. In Chapter4, part context is utilize to label object in clutter. This chapter propose a simultaneous object labeling and localization by using part-whole context for huge 3D objects. Scalability is an important issue in object recognition as it reduces the database storage and recognition time. In this chapter, we propose a new scalable 3D object representation and a learning method to recognize many everyday objects. The key proposal for scalable object representation is to combine the concept of feature sharing with multi-view clustering in part-based object representation, in particular a common-frame constellation model (CFCM). In this representation scheme, we also propose a fully automatic learning method: appearance-based automatic feature clustering and sequential construction of view-tuned CFCMs from labeled multi-views and multiple objects. We evaluated the scalability of the proposed method to COIL-100 DB and applied the learning scheme to 112 objects with 620 training views. Experimental results show the scalable learning results in almost constant recognition performance relative to the number of objects.

5.1 Current state of object recognition

Object recognition has matured, especially in terms of identification level, which local feature-based approaches have made possible. In this chapter, we restrict the scope of object identification to simultaneous object naming and 2D pose estimation. Local features are extracted using the following process: interest point detection [88], characteristic region selection [62], and region description [6, 33, 54, 63]. Based on these studies, several local feature-based object recognition methods, such as probabilistic voting [87], constellation model-based approaches [64], and SVM, AdaBoost [113] have been introduced. The state-of-the-art method SIFT [54] shows very high detection and recognition accuracy in a general environment and is used for categorization [22, 60] and robot localization [89].

However, as the number of objects to be recognized increases, the issue of scalability becomes more important. Conventional object representation methods require linear memory and recognition time proportional to the number of objects. This problem can be serious if the objects are three dimensional, in which storing all the multiple views of 3D objects is impractical.

Partial solutions to the scalability problem have been proposed using image retrieval and indexing. For compact and discriminative feature generation, information theoretic feature selection [108], vocabulary tree generation [73] methods have been proposed. In addition to feature generation, a binary decision tree-based fast indexing method has been proposed [74]. The time taken is proportional to $\log N$, where N is the number of images. Such indexing-based approaches have shown good performance in indexing huge database (DB) of CD titles (in the order of millions) [73] but these methods only focus on image labeling; they cannot handle multiple views of 3D objects and cannot provide pose information, which is very important in object recognition for object manipulation.

Some recent approaches use feature sharing and feature clustering to alleviate the scalability problem. Torralba et al. [101] modified AdaBoost to recognize multi-class objects using a feature-sharing concept. They demonstrated that shared features outperform independently learned features. Murphy-Chutorian and Triesch [68] adopted feature clustering to recognize objects by nearest voting using the clustered-feature DB. Lowe [53] proposed a local feature-based view-clustering scheme to represent multiple views of 3D objects. However, these approaches are only partial solutions that minimize the scalability problem in terms of feature level and multiple view level, respectively.

To reduce the DB size, caused by multiple views of many 3D objects, without degrading recognition performance and motivated by recent studies [53, 68], we propose a new object representation and learning method by combining feature-sharing [68] and view-clustering [53] in part-based object representation [64], as shown in Fig. 5.1.

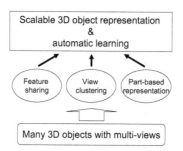

Figure 5.1. Key concept of scalable object representation. Given multi-view images, we build a scalable 3D object representation by automatic learning using feature sharing and view clustering in a part-based representation.

5.2 Related works and scalable 3d object representation

5.2.1 Requirements for scalability and related works

It is evident that simply storing all possible views of many 3D objects requires considerable memory and long recognition time. The main cause of this is the redundancy in DB generation. If we reduce object-related redundancies effectively, we can build a scalable 3D object representation. The necessary requirements for scalable representation are twofold: computational efficiency and low redundancy. Mathematically, an object representation should be as simple as possible, reflecting object variability; and redundant object information should be kept as low as possible to cope with the large number of multi-view objects. We satisfy these requirements by introducing sharing, as shown in Fig. 5.2. We use three types of sharing methods for computational efficiency and redundancy reduction: sharing of view parameters in part-based object representation, sharing of local appearances, and sharing of multiple views.

Sharing of view parameters: We adopt a part-based object representation, specifically a constellation model (CM) [22], instead of a holistic appearance representation [70] because part-based object representation is robust to local photometric and geometric variations, background clutter, and occlusion. We assume an object can

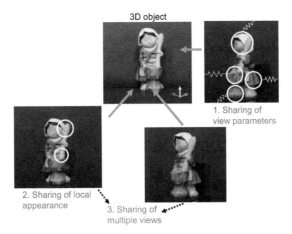

Figure 5.2. Redundancy removal strategy using sharing. Sharing of view parameters in parts reduces computational complexity; sharing of local appearances reduces the number of local features; and sharing of multiple views reduces the number of aspect views.

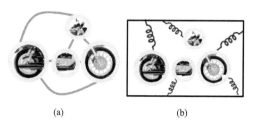

Figure 5.3. Two forms of constellation model: (a) Fully parameterized original constellation model in Gaussian PDF, (b) common-frame constellation model (CFCM) where each part shares object parameters.

be decomposed to a set of visual parts by structure-based approaches (we describe the detail in the following section). There are two forms of CM according to the parameterization method of parts. Fergus et al. [22] proposed a fully parameterized CM, in which appearances and the positions of parts belonging to an object are modeled jointly, or fully parameterized in multi-dimensional Gaussian distributions (see Fig. 5.3(a)). Assume each part is x_i and the number of parts is N, then it can be modeled as a full covariance-based joint PDF, $p(x_1, x_2, \ldots, x_N)$. The degrees of freedom (DOF) of the required parameters are $O(N^2)$. Therefore this model can represent 3 - 6 parts of an object. Instead of this representation, we use a common-frame constellation model (CFCM) representation scheme [64], as this provides advantages in terms of computational efficiency and redundancy removal by sharing object view parameters. If we fix the object ID and viewpoint, each part can share viewing parameters or object frame, $\theta = [objectID,\ pose]$ (see Fig. 5.3(b)). The term "common frame" means the object reference coordinates (shown as the black rectangle in Fig. 5.3(b)), which changes according to the viewing conditions. Then, the mathematical representation can be reduced to the product form, conditioned on an object parameter, such as $\prod_i^N p(x_i|\theta)$. In this scheme, the order is reduced to $O(N)$, which is useful during object learning and recognition. This model can handle hundreds of parts. We refer to this part-based object representation as CFCM because each part shares object parameters (object ID, pose).

Sharing of local appearance: The next redundancy reduction strategy is to share similar part appearances, as shown in Fig. 5.4. It is very inefficient to store all the object images as separate CFCMs as there are many similar parts within an object or shared by different objects. We can remove the redundancies in local appearances

Figure 5.4. Sharing of part appearance for redundancy reduction. The wheels of a bicycle and a motorbike have a similar appearance so they share the same part appearance in CFCMs.

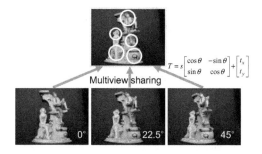

$$T = s \begin{bmatrix} \cos\theta & -\sin\theta \\ \sin\theta & \cos\theta \end{bmatrix} + \begin{bmatrix} t_x \\ t_y \end{bmatrix}$$

Figure 5.5. Sharing of multiple views for redundancy reduction. Multiple views, which are similarity transformed, can be merged into a single CFCM. After view clustering, we can recognize objects from any arbitrary viewpoint.

by sharing similar parts [68,101]. This is a very similar concept to visual codebooks used in the object categorization problem [13].

Sharing of multiple views: In this chapter, we recognize 3D objects from arbitrary viewpoints. Multiple views are used to represent a 3D object. As described above, storing all the training views as separate CFCMs is inefficient. Lowe proposed a local feature-based view clustering method [53] merging multiple training images in the similarity transform space into a single local feature-based model, as shown in Fig. 5.5. We adopt a similar multi-view sharing concept to reduce the number of CFCMs.

5.2.2 Overview of the proposed scalable 3D object representation

Motivated by the three forms of sharing, we propose a new scalable 3D object representation scheme for local feature-based object recognition (simultaneous labeling and pose estimation), as shown in Fig. 5.6. The bottom table shows a library of parts with the corresponding appearance. A local feature used in this work contains an appearance vector and local pose information (region size, dominant orientation, part location relative to reference frame). The appearance of an individual part may be anything, such as a SIFT descriptor [54], moments [63], or PCA [33]. The appearance codebook is generated by clustering of a set of local features extracted from training images. Details are presented in the next section. Each training image is represented by a CFCM where a part has an index (*ID*) to appearance codebook (feature sharing in a CFCM) and local part pose information. Then CFCMs belonging to the similarity transformation space are merged into a single CFCM (multi-view sharing). In the proposed 3D object representation scheme, we use sharing-based redundancy reduction strategies. Parts belonging to an object share an object frame so the complexity of the parameter is a linear function of the number of parts. Because training images are composed of many multiple views and objects, there exist redundant parts and views. We can reduce the redundancies by applying a sharing (clustering) concept to both parts and views. The indicated solid arrow in Fig. 5.6 shows the learned, or view, clustered CFCM.

By means of learning, any 3D object can be represented by a set of view-clustered CFCMs. Each learned CFCM contains object parts that have part pose and link indices to the part libraries (appearance). Likewise, each element in the library contains all the links to the parts in the CFCMs. Note that local appearance codebook libraries are shared between objects and each CFCM contains part poses conditioned on an object frame. The information of part pose is available in [54,88] and this is important for bottom-up object recognition and verification. We can use this fact to generate hypotheses during object recognition. The next two sections describe the details of learning by feature and by view-clustering respectively.

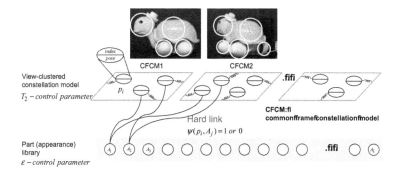

Figure 5.6. Scalable 3D object representation scheme. The local appearance codebook is shared between objects, part parameters are shared in a CFCM, and multiple views are clustered to a CFCM

5.3   Visual feature and automatic clustering

5.3.1   Local feature library generation by appearance-based automatic feature clustering

Basically, we utilize G-RIF as a feature detector introduced in Chapter 3 for its robustness to visual variations [35]. A local appearance library or codebook can be generated by clustering training features. There are several clustering methods, such as the k-means algorithm, vector quantization [15], and mean-shift [26]. The k-means algorithm and vector quantization methods are based on iterative optimization starting from random cluster centers with a predetermined number of clusters. Mean-shift is a superior method of mode estimation for dense features, because it is based on derivatives of feature density but, experimentally, it did not show good results for our high dimensional sparse feature set, where the dimension of a feature was over one hundred (189) and the size of a feature was more than several hundreds of thousands. In addition, the conventional energy minimization-based approach, such as k-means clustering, is difficult to use due to slow convergence.

The main problems of the k-means algorithm for high-dimensional data are:
1) How to set the cluster size (k).
2) How to set the initial cluster centers.

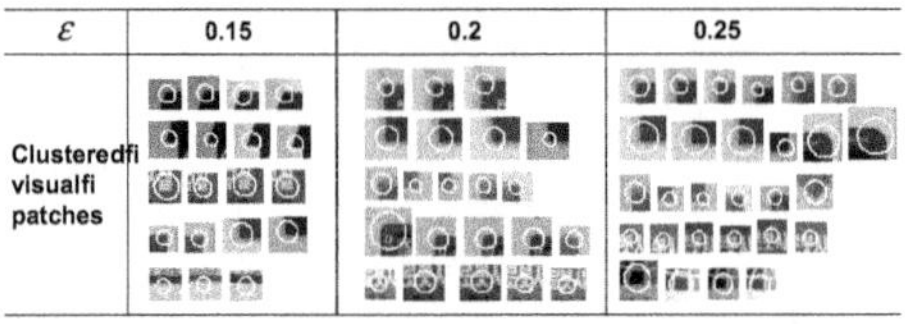

ε	0.15	0.2	0.25
Clusteredfi visualfi patches			

Figure 5.7. Appearance similarity threshold (ϵ) and corresponding part clustering results. Each row represents clustered parts. As the similarity threshold becomes larger, fewer similar parts are clustered.

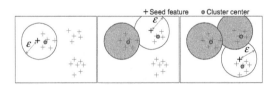

Figure 5.8. Automatic initial clustering procedures: (left) random selection and ϵ-nearest clustering, (middle) after removing the clustered features, process (left) is iterated, (right) clustering-removal is iterated.

3) How to effectively compare distances between data and cluster centers. We propose a simple but practical clustering algorithm suitable for high dimensional visual features. We solve the above problems by using the properties of appearance and a nearest-neighbor search using a k-d tree [3]. Since feature descriptors are energy-normalized vectors, the Euclidean vector distance can provide the appearance similarity. Fig. 5.7 shows part clustering results for the Euclidean distance threshold (ϵ). The circle represents a detected part using G-RIF. We can cluster visually similar parts using only one parameter, ϵ. As the threshold becomes larger, approximately similar structures are clustered. Note that we can determine the number of clusters automatically if we use the visual distance threshold.

In part-based object recognition, part structures have very significant roles. Consequently, first we find approximate structure centers by sequentially performing the ϵ-nearest neighbor search, as shown in Fig. 5.8. The clustered features are removed in search space. Then, cluster centers are optimized using k-means clustering. This process corrects features on the cluster boundaries. Means and variances of the appearance vectors are stored for inference. We assume diagonal covariance with equal variance for simplicity. Algorithm 5.1 summarizes the whole clustering steps. By merging the ϵ-nearest neighbor search, k-d tree-based distance calculation, and k-means algorithm, we can solve the three problems simultaneously. The number of clusters is estimated by ϵ. The ϵ-nearest neighbor-based approximate clustering provides a suitable starting point for k-means clustering. In addition, a k d tree-based data search reduces the search time for large volumes of data.

Fig. 5.9 shows the convergence rate of clustering with the proposed initialization and random samples in k-means clustering for 72,083 features extracted from 620 images . Automatic clustering almost converged within two iterations. This can be explained by the good initial estimation of cluster centers using sequential approximate clustering. The conventional k-means algorithm converges more slowly than the proposed algorithm.

5.4 Learning scalable object representation by sequential construction of shared feature-based view clustering

We represent a 3D object by a set of view-tuned CFCMs. Visual parts in a CFCM are conditioned by the view-tuned parameters. The term view-tuned means view

---

Algorithm 5.1 Automatic clustering of high-dimensional vector

Given: training images

Step 1. Extract all features (G-RIF) from multi-views and multi-objects.

Step 2. Make a k-d tree of training features.

Step 3. Do

 For all training features

  Randomly select a feature (bold cross in Fig. 5.8(left))

  Find ϵ-nearest features from search DB (features within a circle in Fig. 5.8(left)).

  If number of clustered features >0

   Calculate cluster center of ϵ-nearest features (small dot in Fig. 5.8)

   Delete ϵ-nearest features in search DB (shaded circle in Fig. 5.8)

  End if

 End for

Step 4. Do the optimization using k-means clustering starting from the cluster centers.

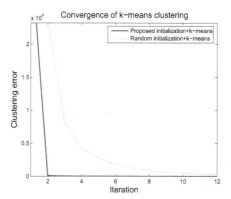

Figure 5.9. Convergence comparisons between the proposed automatic clustering and the conventional k-means algorithm.

clustering in a similarity transformation space or reference object frame. Fig. 5.10 shows the overall object learning structure. Given labeled multi-views and multi-object images, we find view-tuned CFCMs. In a CFCM, each part is represented in terms of local pose and appearance index of the shared feature libraries learned in the previous section.

Learning is conducted sequentially; we set the first image as the reference CFCM. As shown in Fig. 5.11, a CFCM is generated from detected features and the feature library, it contains object ID, view ID, and parts (pose, appearance per part). The pose of a part is obtained directly from the feature detector (part size, dominant orientation, image location to the reference frame), and the part appearance is represented using the index to the shared appearance library. Since we only have a CFCM, positional uncertainty (σ_{loc}) is set to a default value. The uncertainty is updated using the following process.

From the next image, we extract local features using G-RIF detectors. Matching pairs are searched for between the input feature points and the part locations in CFCMs, using the appearance distances between input feature descriptors and appearance feature library within ϵ-nearest neighbors. Then we perform a Hough transform in pose space (CFCM ID, scale, orientation space), as shown in Fig. 5.12(b).

Finally, new CFCMs are constructed, or previously learned CFCMs are updated,

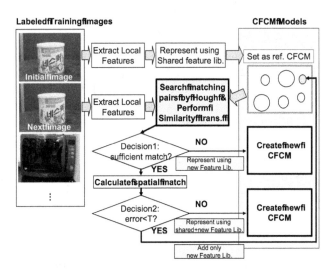

Figure 5.10. Object learning by the sequential view clustering. Given an appearance library, image features are extracted from each training image. Then proper clustering action (Case I, II, III) is selected based on the result of core functional blocks

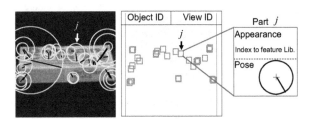

Figure 5.11. CFCM representation from detected features: (Left) object decomposition into parts using G-RIF (middle) A CFCM represents object ID, view ID, part information relative to the reference frame, (right) close up view of part j.

in one of the following three cases.

CASE 1: Completely different objects or views

If the number of matching pairs between an input image and a candidate CFCM is below a predefined threshold ($T1 = 5$), then it generates a new CFCM, represented in terms of new feature libraries as shown in Fig. 5.11.

CASE 2: Similar views but with large similarity transform error

If there are sufficient matching pairs but the spatial average matching error by similarity transformation (see Fig. 5.12(d)) is larger than a predefined threshold (usually $T2 = 20$ pixels, validation in experimental section), then a new CFCM with a shared-feature and with new feature libraries is created, as shown in Fig. 5.11. The CFCM generation procedure is the same as in CASE 1 but the number of shared features in CASE 2 is relatively higher than that for feature sharing in CASE 1.

CASE 3: Similar views with small similarity transform error

Finally, if the matching between input and candidate CFCM is almost correct (number of matching $> T1$ and similarity transform error $< T2$), then we add distinctive new features to the existing CFCM, as shown in Fig. 5.10 (green circle in CFCM). The distinctive features can enhance the discriminative power during inference. We define a new input feature (f_{New}) as distinctive if it is very close to the shared feature library (f_{Lib}) and the sharing number of the library feature ($N_{Shared}^{f_{Lib}}$) is as low as possible to reduce ambiguity, it is expressed in Eq. 5.1. In CASE 3, pose information of parts in a CFCM is updated from overlapping part pairs between the input and the candidate CFCM. For simplicity we update means and variances of part locations.

$$DM(f_{New}, f_{Lib}) = \frac{exp(-N_{Shared}^{f_{Lib}})}{dist(f_{New}, f_{Lib})} \tag{5.1}$$

We applied the sequential learning method to the COIL-100 DB [71], which is available at [http://www1.cs.columbia.edu/CAVE/software/softlib/coil-100.php]. Fig. 5.13 shows an example of sequential CFCM construction results from full 3D views of an object. We obtained 14 view-tuned CFCM images from 36 training images. The CFCM construction method can extract distinguishable multiple views for 3D objects in similarity transformation space (affine transformation is not suitable for 3D objects since the feature detector is invariant under the similarity transform).

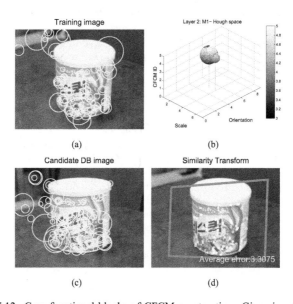

Figure 5.12. Core functional blocks of CFCM construction. Given input images with detected features (a), similar CFCM (c) is found using Hough transform (b). Similarity transform error is calculated between an input image and a candidate CFCM.

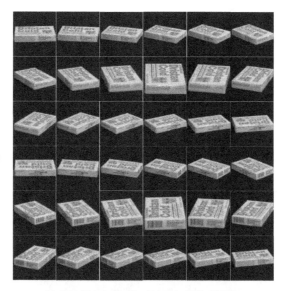

(a) 36 training images for object 1 in COIL-100 DB

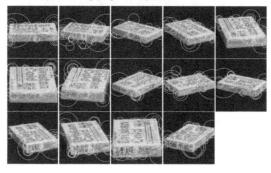

(b) 14 learned CFCMs

Figure 5.13. CFCM construction example. The sequential construction of CFCM can compress 36 training images into 14 CFCMs (compression rate: 61.1%)

5.5 Multi-object recognition

We can fully utilize the scalable object representation by the shared feature-based view clustering method in object recognition by using the well-known hypothesis and verification framework. However, we modify it to recognize multiple objects in the proposed object representation scheme.

If S represents a set of scene features, D represents a set of database entries (shared feature lib. + CFCMs), and H represents hypothesized CFCMs, which best describe the scene, then the object recognition problem can be formulated as a mixture form (the 1st line in Eq. 5.2: assuming multiple objects in a scene). π_m is the mixture weight of object m which is estimated on-line by a set of CFCMs belonging to m. $\hat{h}_m$ is the optimal transformed CFCM for object m. If we assume uniform priors, the equation can be reduced to the second line in Eq. 5.2. We select the best hypothesis ($\hat{h}_m$) which has the maximum conditional probability ($p_m(S_m|h_m^{(i)}, D)$).

$$
\begin{aligned}
p(H|S, D) &= \sum_{m=1}^{M} \pi_m p_m(\hat{h}_m|S_m, D) \\
&\propto \sum_{m=1}^{M} \pi_m p_m(S_m|\hat{h}_m, D)
\end{aligned}
\tag{5.2}
$$

where $p_m(S_m|\hat{h}_m, D) = \arg\max_{i \in I_m}\{p_m(S_m|h_m^{(i)}, D)\}$, $\sum_{m=1}^{M} \pi_m = 1$. We can model $p_m(S_m|h_m^{(i)}, D)$ by a Gaussian noise model of appearance and pose using Eq. 5.3. We assume that the appearance and pose of each part is independent. In addition, since features in a CFCM are conditioned on a common-frame, they can be handled independently. $\mathbf{y}_{app}$ is the shared feature closest to scene feature $\mathbf{x}_{app}$. $\mathbf{y}_{loc}$ is the position of a part hypothesized by $h_m^{(i)}$. σ_{app}, σ_{loc} are estimated from training data, as described in previous sections.

$$
p_m(S_m|h_m^{(i)}, D) = \prod_{\mathbf{x} \in S_m} p_{app}(\mathbf{x}|h_m^{(i)}, D) \cdot p_{pose}(\mathbf{x}|h_m^{(i)}, D)
\tag{5.3}
$$

where $p_{app}(\mathbf{x}|h_m^{(i)}, D) \propto exp(-\|\mathbf{x}_{app} - \mathbf{y}_{app}\|^2 / \sigma_{app}^2)$, $p_{pose}(\mathbf{x}|h_m^{(i)}, D) \propto exp(-\|\mathbf{x}_{loc} - \mathbf{y}_{loc}\|^2 / \sigma_{loc}^2)$.

Fig. 5.14 summarizes the object recognition procedures graphically. We can obtain all possible matching pairs by an ϵ-NN (nearest neighbor) search in the feature library. Because each feature library contains all possible links to CFCMs, we can form all the matching pairs between input image features and parts in CFCMs. From these, hypotheses are generated by the Hough transform in the CFCM ID, scale (11 bins), orientation (8 bins) space [54], and grouped by object ID. Colors

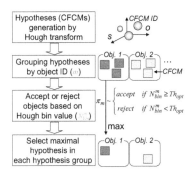

Figure 5.14. Hypothesis and test (verification)-based object recognition procedures.

in Fig. 5.15(b) represent object groups. Each Hough particle (hypothesis) in Fig. 5.15(b) is shown in Fig. 5.15(c). Then we decide whether to accept or reject the hypothesized object based on the bin size with an optimal threshold [68]. Finally, we select optimal hypotheses, using Eq. 5.3, which can best be matched to object features in a scene. Fine object poses are refined by similarity transformation between input image features and the selected CFCM.

5.6   Evaluation of scalability

We evaluated the proposed method to both COIL-100 DB [71] and a dataset captured in a cluttered indoor environment. The COIL-100 DB consists of 100 3D objects with 72 multiple views per object acquired in a controlled environment; this public DB is suitable for evaluating the scalability of the proposed algorithm. Our DB was acquired at arbitrary viewpoints in a cluttered indoor environment and it also contains multiple objects per image. This DB was used to evaluate the recognition performance in a real environment.

5.6.1   Proposed object representation vs. baseline method

To validate the scalability of the proposed method, we compare the proposed object representation with an object recognition method without sharing (we call it

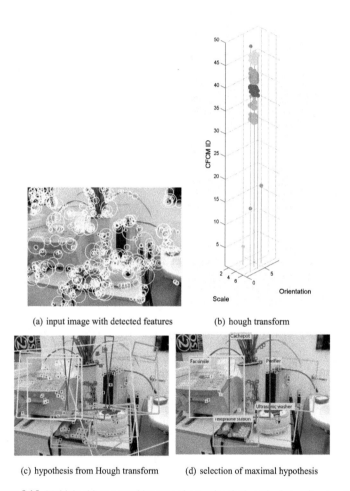

(a) input image with detected features   (b) hough transform

(c) hypothesis from Hough transform   (d) selection of maximal hypothesis

Figure 5.15. Multiple object recognition procedures using the hypothesis-test framework.

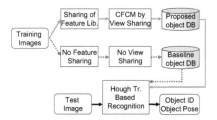

Figure 5.16. Configuration of object representation and test. Solid red arrows represent the proposed object learning procedures and dotted blue arrows represent baseline object representation. The same object recognition method was used for performance comparison.

a baseline method in this chapter, as shown in Fig. 5.16. In the baseline object representation, we did not conduct the feature sharing and view sharing process. We directly stored the detected features (part appearance + part pose) of each image as separate object models. This is a conventional object modeling method used in [54,64,87]. If the number of object is 100 and each object has 36 multiple views, then the total number of object models is 3,600 in the baseline method. Both object representation schemes used the same object recognition method described in previous section for fair comparison of scalability.

5.6.2   Evaluation on COIL-100 DB

COIL-100 DB contains 100 man-made objects with 72 views per object. Fig. 5.13(a) shows partial multiple view images for object 1. Since the objects were acquired in a controlled environment and have multiple views for 3D objects, we can evaluate scalability. We divided the multiple view images (72 views/object) into two sets for training (36 views/object) and for testing (36 views/object).

Parameter selection: First, we have to set two parameters, such as ϵ, $T2$. As described in section 2, ϵ can control the size of the feature library. Likewise, $T2$ can control the size of the CFCMs per object. Optimal values are selected when the recognition rate is high enough. In this test, we used 10 3D objects with 360 multiple views. Fig. 5.17 shows the evaluation results for threshold ϵ. In this test, we fixed $T2$ as 15 [pixels] and varied ϵ from 0.1 to 0.5. Ideally, the size of the shared features should be as small as possible. At the same time the shared feature library should give a high object recognition rate. From the graphs shown in Fig.

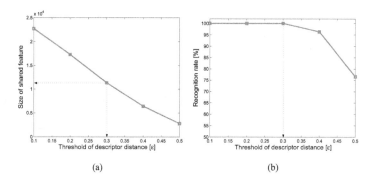

Figure 5.17. Size of shared feature (a) and recognition rate (b) for varying ϵ (distance threshold used to measure descriptor similarity). As the threshold becomes larger, the size of the shared feature reduced in inverse proportion. Recognition rate drops abruptly at ϵ

5.17, the optimal parameter selection for ϵ is 0.3.

The parameter $T2$ was selected in a similar manner to the above evaluation procedure. Shared features were generated using fixed ϵ, then various $T2$ thresholds provide different sizes of CFCMs. When $T2 = 1$, the average number of CFCMs per object is 35 (351/10) with a recognition rate of 100.0% as shown in Fig. 5.18. In this case, there is almost no view clustering (no compression) with successful recognition performance. When $T2 = 30$, the average number of CFCMs per object is 8 (80/10) with a recognition rate of 98.8%. Although this parameter value provides strong view clustering (36 → 8 /object), the recognition performance is relatively low due to incorrect pose estimation (we used similarity transformation in CFCM construction). The best selection for $T2$ is 20 [pixels] where we obtained a sufficiently high recognition rate (99.7%) with 10 CFCMs per object (sufficient compression). In the following scalability experiments, we used $\epsilon = 0.3, T2 = 20$, for feature sharing and view clustering, respectively.

Scalability test: To validate the scalability of the proposed object representation method, we compared DB size, recognition time, and recognition rate between the baseline method and the proposed method by changing the number of objects; we used 20, 40, 60, 80, and 100 objects. Each object had 36 multiple views. Object models of the baseline method were generated by setting $\epsilon = 0.0, T2 = 0$, (no compression). The proposed object models were generated using parameters $\epsilon =$

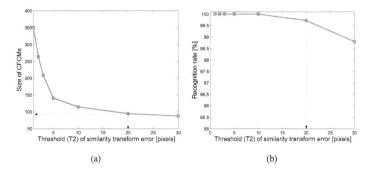

(a)   (b)

Figure 5.18. Size of CFCMs and recognition rate for varying $T2$ (threshold of similarity transform between input reference frame and corresponding CFCM). (a) As the threshold increases, the number of multiple views per object is reduced in inverse proportion. (b) The recognition rate drops abruptly at $T2 = 20$.

$0.3, T2 = 20$. Fig. 5.19 summarizes the overall performance evaluation in terms of model size (number of shared features, number of CFCMs), recognition time, and recognition rate.

The model size and time complexity of the baseline method are linearly related to the number of objects. In this case the recognition rate is high because it uses all the training images. However, the proposed method have a lower than linear relation to the number of objects. Note that the recognition rate at the object size 20 is 99.6% and it only drops 1.2% remaining at over 98.4% for an object size of 100. These graphs validate the scalable properties of the proposed object representation scheme. Note that the recognition performance of our scalable object recognition system is comparable to a state-of-the art indexing method [74]. The proposed method is 98.4% accurate and the indexing-based method is 99.8% for the same COIL-100 DB but our system can provide pose information in addition to an object label. The average recognition time for 100 objects (image size: 128×128) is only 4sec compared to 14sec in the baseline method with a AMD 4800+ machine. Fig. 5.20 shows several successful (correct name + correct pose) object recognition results. Most failures originated from a very small number of local features and very similar objects (objects 29 and 32 shown in Fig. 5.21). In this situation, we cannot calculate the correct object pose.

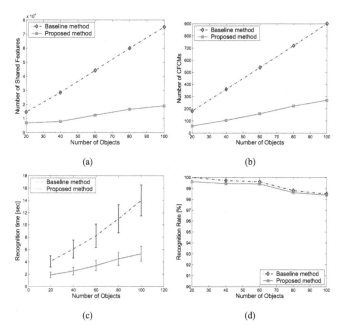

Figure 5.19. Performance evaluation on COIL-100 DB. We compare the proposed object representation method to the baseline method. (a) Number of shared features vs. number of objects, (b) Number of CFCMs vs. number of objects, (c) Recognition time vs. number of objects, and (d) Recognition rate vs. number of objects.

Figure 5.20. Examples of successful object recognition using the proposed object representation and recognition method for COIL-100 DB.

Figure 5.21. Failure recognition due to very similar objects. The two left-hand images represent object models and the right-hand image shows the object recognition result. Our recognition method shows strong evidence of object 29, 32 in this case.

Figure 5.22. Examples of labeled multi-view/multi-object images for training.

5.6.3 Evaluation on the real data set

We prepared 112 objects to test the scalability of the proposed method in a general environment where many objects exist, with clutter. The segmented objects for training are labeled as shown in Fig. 5.22. The total number of training object images was 620. We used 228 test scenes where each scene contained zero to six objects (total: 247 objects) not used in the learning process (different camera pose and lighting conditions). The size of test image is 640×480.

Given the huge training data set, we first extracted all visual features using G-RIF. Then we applied automatic clustering using ϵ. Based on these clustered features, we sequentially constructed view-tuned CFCMs using the learning method shown in Fig. 5.10. The size of the view-tuned CFCM was determined by the threshold ($T2$) of similarity transform error (pixel). Table 5.1 summarizes these learning results by changing two parameters. We reduced the number of features by feature sharing and also reduced the size of the CFCMs by view clustering.

Next, we evaluated the recognition performance for the range of learning data shown in Table 5.1. We decided that a recognition result is successful if both the object ID and pose are correct when evaluated by the human eye. Fig. 5.24 shows the results. Fig. 5.23(a) is obtained by fixing the size of the CFCM at 264 ($T2 = 20$). As the feature size increases, the system achieves higher recognition rates. However, the recognition almost converged for the size of feature 35,314

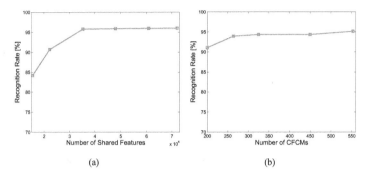

(a)   (b)

Figure 5.23. Evaluation of object recognition for different DB sizes: (a) number of shared features (b) number of view-tuned CFCMs.

($\epsilon = 0.3, 95.8\%$). Likewise, we can obtain the recognition performance according to the size of a view-tuned CFCM (we fixed the number of features as 35,314), as shown in Fig. 5.23(b). We can achieve very high accuracy with only 264 CFCMs ($T2 = 20, 95.8\%$).

Finally, we checked the recognition time and recognition rate for varying numbers of objects. Note that the recognition time is log-linear with respect to the number of objects, as shown in Fig. 5.24(a). Furthermore, the recognition rate is almost constant irrespective of the number of objects, as shown in Fig. 5.24(b).

From these experiments, we can deduce that the proposed object representation scheme is scalable. If we set ϵ to 0.3, $T2$ to 20, then the overall recognition rate is

Table 5.1. The size of clustered features and CFCMs is reduced as the thresholds (ε, $T2$ respectively) increase.

ε	0	0.1	0.2	**0.3**	0.4	0.5
No. of shared feature	72,083	66,081	48,063	**35,314**	22,183	15,416
$T2$	0	1	4	10	**20**	30
No. of view-tuned CFCM	620	553	450	325	**264**	200

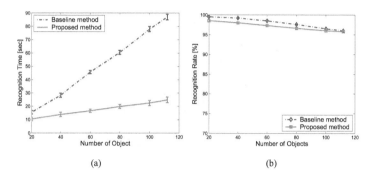

Figure 5.24. (a) Recognition time vs. number of objects, (b) recognition rate vs. number of objects.

95.8%. Fig. 5.25 shows examples of multiple object recognition results by selecting maximal CFCMs in multi-modal probability, using Eq. 5.2.

5.7 Summary

In this chapter, we proposed object identification and localization using part-object context. Especially, we focus on scalable 3D object representation and learning by combining feature sharing (part) and view clustering (object) in part-based object recognition. First, we decompose an object into convex and corner parts described by a localized histogram of edge density, orientation and hue. Visual structure-based automatic clustering is particularly useful for feature sharing and sequential construction of CFCMs, which can learn any new incoming objects of practical importance. We experimentally validated that the shared feature-based view-clustering scheme can effectively represent 3D objects and is scalable to the number of objects. We recognize multiple objects by a hypothesis and verification method in identification levels. In Chapter 8 and 9, we will investigate upgrading the scalable object representation and learning scheme to the categorization level by extending the concept of a codebook. We will also model object context and scene context for more robust scene interpretation using a hierarchical graphical model in Chapter 6.

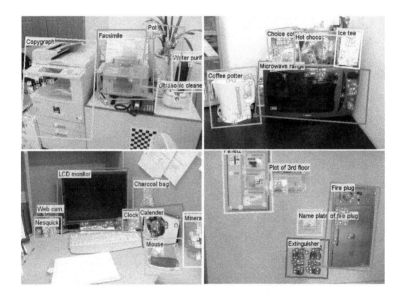

Figure 5.25. Object recognition results by a hypothesis and verification scheme. The proposed multiple object recognition method can detect multiple objects in various indoor environments.

CHAPTER VI

SPATIAL AND HIERARCHICAL CONTEXT: SCENE INTERPRETATION

Partial contexts such as pixel, part, part-object are introduced in previous Chapters. Based on this work, we present a new system for collaborative place, object, and part recognition in indoor environments by integrating spatial and hierarchical context. We consider a scene to be an undirected graphical model composed of a place node, object nodes, and part nodes with undirected links. Our key contribution is the introduction of collaborative place and object recognition (we call it as the hierarchical context in this chapter) instead of object only or causal relation of place to objects. We unify the hierarchical context and the well-known spatial context into a complete hierarchical graphical model (HGM) (partial contexts are introduced in previous chapters). In the HGM, object and part nodes contain labels and related pose information instead of only a label for robust inference of objects. The most difficult problems of the HGM are learning and inferring variable graph structures. We learn the HGM piecewise instead of by joint graph learning for tractability. Since the inference includes variable structure estimation with marginal distribution of each node, we approximate the pseudo-likelihood of marginal distribution using multi-modal sequential Monte Carlo with weight update by belief propagation. Data-driven multi-modal hypothesis and context-based pruning provide the correct inference. For successful recognition, issues related to 3D object recognition are also considered and several state-of-the-art methods are incorporated. The proposed system greatly reduces false alarms using the spatial and hierarchical contexts. We demonstrate the feasibility of the HGM-based collaborative place, object, and part recognition in actual large-scale environments for guidance applications (12 places, 112 3D objects).

Figure 6.1. Our place and object recognition system can guide visitors with a wearable camera system. A USB camera carried on the head can provide image data to a notebook computer in a carry bag. Processed information is delivered to the visitors as audiovisual data.

6.1 Place and object recognition

Consider visitors to a new environment. They have no prior information about the environment so they may need a guidance system to acquire the place and related object information. This chapter is concerned with the problem of recognizing places and objects in real environments, as shown in Fig. 6.1. The scope of the place and object recognition is to identify places and objects with parts in the form of place labels and object labels with poses. This can be regarded as scene interpretation at the semantic level. It is fundamentally important to recognize places and objects in uncontrolled environments where the camera may move arbitrarily and light conditions also change.

We are aware of only two approaches to place and object recognition. One baseline method regards these as separate problems. The other method directly relates places and objects using a Bayesian framework [102] (dotted arrow as shown in Fig. 6.2). Place is recognized first using gist information from filter responses and then the place information provides the Bayesian probable prior distribution of object label and scale, position [102]. The first contribution of this chapter is the proposition of an interrelated place, object, and part recognition method using an undirected graphical model [32]. Place information can provide contextually related object priors, but conversely, ambiguous places can be discriminated by recognizing contextually related objects. This is the key concept for collaborative place and object recognition. Likewise, object information can provide contextually related part priors, but conversely, parts can provide evidence for the existence of a

Figure 6.2. Previous approaches regard place recognition and object recognition as separate problems or directly (causally) related problems. Our approach (solid line) regards them as interrelated problems.

specific object. This work is motivated from the recent neurophysiological findings by Bar [4], in which objects are strongly related to specific scenes or places. A drier is strongly related to a bathroom and a drill is strongly related to a workshop and so on.

As a second contribution, we extend the concept of interrelation between place and object to the object and part level so that we unify hierarchical context and the well-known spatial context [5] in objects and parts into a hierarchical graphical model (HGM), as shown in Fig. 6.3. Originally, the term "context" was used in linguistics to represent verbal meanings from relationships within a sentence. We use the term throughout this chapter to represent information coming from "visual relationships" in images. There is one place node, an object layer composed of multiple object nodes, and a part layer composed of multiple part nodes. In this HGM, the bidirectional interaction properties of nodes within layers (dotted lines in Fig. 6.3(b)), and nodes between layers (solid line in Fig. 6.3(b)) are important. A third contribution of this work is the proposition of approximated learning by a piecewise method instead of holistic minimization of joint likelihood for tractable learning. A fourth contribution is the simultaneous inference of variable structure and marginal distribution using multi-modal sequential Monte Carlo method (MM-SMC) with weight calculation by belief propagation (BP).

As a final contribution, the fundamental issues (see Fig. 6.4) related to 3D objects are considered and a robust and scalable 3D object representation (introduced in Chapter 5) is incorporated in the nodes of HGM. The issue of figure/ground segmentation is bypassed by using semantic local features and using background information in the form of visual context solves the issue of background clutter.

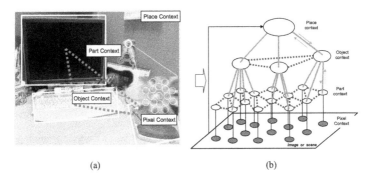

(a)   (b)

Figure 6.3. Unification of visual context: (a) A scene contains place, objects, parts, and
pixel information, and (b) they are integrated into a hierarchical graphical model (HGM).
Red solid lines represent hierarchical contextual relationships, blue dotted lines represent
the well-known spatial context, and black lines represent measurements.

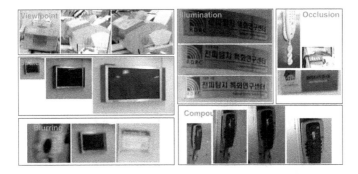

Figure 6.4. Fundamental issues in 3D object recognition are visual variations with changing viewpoint, scale, illumination, occlusion and blurring. These issues arise concurrently
in working environments.

6.2   Related Works

In this section, we introduce related work focusing on place and object recognition in real environments. There are two approaches for place and object recognition, according to whether the visual relationships are modeled or not.

6.2.1   Independent approach to place and object recognition

Place recognition or scene classification is an active research area. Vogel and Schiele proposed a natural scene retrieval system based on a semantic modeling step [112]. They classify image regions into semantic labels, such as "grass" and "rocks". Co-occurrence vectors of labels can provide scene categories of coasts and rivers, for example. Fei-Fei and Perona applied latent semantic analysis in a text categorization technique to the scene categorization problem and obtained satisfactory results for 13 categories of office, kitchen, street, etc. [21]. However, the place recognition researchers did not consider any object recognition information, so place recognition can be ambiguous in similar environments.

In the object recognition problem, local feature-based approaches show groundbreaking performance for textured objects [6, 54, 62, 63, 88]. Of these, SIFT (scale invariant feature transform) feature, proposed by Lowe, shows robust performance in object recognition [54]. This feature can cope with partial view changes, scale changes, illumination changes, and occlusions, as shown in Fig. 6.4. However, local feature-based object recognition cannot discriminate between ambiguous objects.

6.2.2   Graphical model for visual relation

Unlike independent place and object modeling, there are several modeling methods, depending on type, for visual relationships. Basically, we can categorize visual relations into spatial contexts and hierarchical contexts. The first is visual interaction in image space, such as pixels, parts, and objects. The second is the interaction between abstraction levels, such as place-objects and object-parts. Many researchers have mathematically modeled the relational information, using a semantic network [72] and a graphical model [32]. A semantic network can represent concepts in declarative knowledge and inference is performed using rule-based procedural knowledge. Although the semantic network intuitively provides a method suitable for our problem, we adopt a graphical model because it can provide a principled method for the modeling of visual relations (graph theory) and uncertainty (probability theory). The graphical model is the combination of probability theory and graph theory. We can subdivide the previous work on graphical models according to which of five conditions it uses: single layer, multi-layer, linking method, graph structure, and node information, as shown in Fig. 6.5(a).

Single layer: Most undirected graphical models (noncausal models) have a single layer to describe spatial context. There are generative models of the Markov Random Field (MRF) [49] and discriminative models of the Conditional/Discriminative Random Field (CRF or DRF) [40]. Recently, CRF-related methods have been proposed, such as Mutual Boosting [23] and Boosted Random Field (BRF) [101], for easy learning and discriminative inference of hidden labels.

Multi-layer: Multiple hidden layers can be used to incorporate larger spatial interactions by multi-resolution [29, 100], or through different semantic abstraction levels, such as scenes, objects, and parts [41, 67, 102].

Linking method: The most important and difficult element in multilayered graphical models is the linkage of the multiple layers. The simplest method is to produce each layer (fully undirected model) where there is the same type and number of nodes in each layer, such as mCRF using Expert-of-Product method [29]. Other popular methods are directly linked methods. One is the top-down Bayesian Network, or generative model [67, 97, 100, 102], and the other is the bottom-up, or discriminative, model, such as the Hierarchical Random Field (HRF) formed by directly linking two CRFs [41].

Graph structure: The graphical models above assume a fixed graphical model, which is rather a simple problem because only the marginal distribution of each node has to be estimated. If the graph structure varies from image to image then this becomes a very challenging problem. Either the number of nodes is fixed and only the links can be variable, such as dynamic trees [96], or both nodes and links can be variable [34, 103]. For variable node estimation, trans-dimensional Markov Chain Monte Carlo (TD-MCMC) is frequently used since mathematical convergence is guaranteed [103].

Node information: Traditionally, most graphical models only estimate labels as hidden variables. Recently, position information is encoded into dynamic trees and this shows better performance for image labeling [96]. In DRF, domain information of the patch location is reflected in the label interaction [40]. We extend the location information to object pose and part pose for accurate inference.

In our problem, we use three layers composed for place, object, and part. Each layer is conditionally linked. The place node contains only the place label. Part node and object node hvae the labels with poses. Therefore, each layer has a pose-encoded-MRF structure, as shown in Fig. 6.5(b).

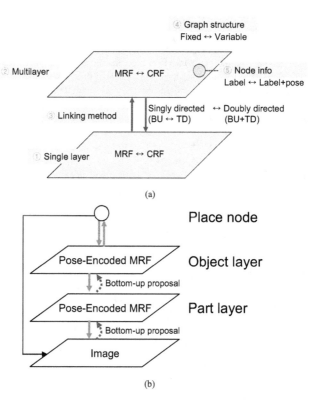

(a)

(b)

Figure 6.5. (a) Aspects of graphical modeling for visual relations. The graphical model can be divided into single layer or multi-layer, by the linking method between layers, graph structure, and node information, (b) our proposed scheme consists of multi-layers, doubly directed linking and variable graph with pose-encoded-MRFs for collaborative place, object, and part recognition.

6.3   Hierarchical Graphical Model for Spatial and Hierarchical Context

Consider the real computer desk scene shown in Fig. 6.3(a). We wish to recognize place and objects with parts from the image. A desk provides the prior of contextually related objects (hierarchical context); a desk usually contains computer sets. All objects are spatially correlated (spatial context); a monitor, keyboard, and mouse co-occur on a desk. Each object is also composed of spatially related visual parts (hierarchical context); a mouse has buttons and a pad. These visual parts are spatially related (spatial context); the mouse buttons and pad coexist.

The problem is to model such visual context for successful place and object recognition. As described in previous section, direct extension of the simple Bayesian formula proposed by Torralba [102] may be a starting point. However, this approach cannot model important contextual properties, such as the spatial and hierarchical interaction. Markov Random Field (MRF) or Discriminative Random Field (DRF), which are frequently used for image segmentation, may be more suitable for modeling spatial interaction [40, 49]. Kumar and Hebert extended DRF to two layers that are directly linked for longer contextual influence in segmentation [41]. Tu et al. proposed an image parsing method using bottom-up/top-down information with a Monte Carlo Markov Chain framework [103]. However, this did not explicitly model the spatial interaction of visual objects or parts. In this section, we propose an HGM to reflect spatial and hierarchical visual contexts simultaneously at an identification level. Identification level means that we deal with specific scenes that include specific objects in a certain working environment.

6.3.1   Mathematical formulation for place, object, and part recognition

Although the directed graphical model of a Bayesian network may be computationally feasible, causal relationships between visual objects are undesirable, since the failure of one component leads to a cascading failure to recognize other objects [18]. Conceptually, the spatial context and the hierarchical context for place, object, and part recognition can be modeled as an undirected graphical model, as shown in Fig. 6.3(b). A graphical model is a useful tool because it can model a complex system by a set of probabilistic modules in [32]. Nodes represent random variables and arcs represent probabilistic interactions between nodes. White nodes represent hidden

variables and black nodes represent visual observations. The arcs between hidden
nodes are represented by compatibility or correlation functions. Thick solid arcs
represent the hierarchical context and dotted arcs represent the spatial context in
Fig. 6.3(b). For example, the detection of each of monitor, keyboard, and mouse
reinforces one another, which leads to stable recognition. In the HGM, contextual
facilitation mechanisms are reflected everywhere. A scene node is related to the object nodes and the object nodes are also related to the part nodes. Scene and objects
are processed in parallel (thin solid lines). The scene node receives evidence from
holistic image features and each part receives evidence from an individual image
feature. The black nodes represent visual features extracted from an image.

Let $x^S \in \{0, 1, 2, \cdots, N_S\}$ be a place label where 0 is the unknown place, $x^O = \{n_O, \theta_O\}$ is the object label ($n_O \in \{1, 2, \cdots, N_O\}$) with pose ($\theta_O$: similarity transform from model to image), and $x_P = \{n_P, \theta_P\}$ is the part label ($n_n\{0, 1, 2, \cdots, N_O\}$
with pose (observed)). The pose information is very important for successful ob-
ject recognition. Note that most graph-based methods use only label information [29, 100]. Since the objective of this work is to recognize a place and multiple
objects with related parts from a given image, the ideal inference can be performed
similar to maximizing a posteriori probability (MAP) as :

$$\{\hat{x}^S, \hat{\mathbf{x}}^O, \hat{\mathbf{x}}^P\} = \arg\max_{x^S, \mathbf{x}^O, \mathbf{x}^P} p(x^S, \mathbf{x}^O, \mathbf{x}^P | \mathbf{y}) \tag{6.1}$$

If the graph structure is known (fixed), and all the compatibilities are given,
we can estimate the joint probability distribution from Eq. 6.2, where ψ denotes
the compatibility, or correlation, functions between hidden nodes and ϕ denotes the
compatibility between the hidden nodes and the observation nodes.

$$p(x^S, \mathbf{x}^O, \mathbf{x}^P | \mathbf{y}) = \frac{1}{Z} \phi(x^S, \mathbf{y}) \prod_i \psi(x_i^O, x^S) \prod_{(i,j)} \psi(x_i^O, x_j^O) \prod_{(i,j)} \psi(x_i^P, x_j^O) \prod_{(i,j)} \psi(x_i^P, x_j^P) \prod_{(i)} \phi(x$$
$$\tag{6.2}$$

The scene layer or place node receives evidence from image features and objects. The object layer receives information from parts, a scene and neighboring
objects. The part layer receives evidence from features, objects and neighboring
parts. However, there are two computational problems in Eq. 6.2. First, the computation of the global normalization factor or partition function Z is intractable due
to its complexity, and second, the graph structure is not fixed during learning and
inference because we do not know the number of object nodes before they are manually given or recognized. The first problem caused by the global partition function

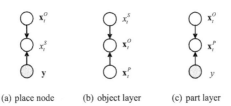

(a) place node   (b) object layer   (c) part layer

Figure 6.6. Pseudo-likelihood approximation of the original undirected graphical model for tractability of the global partition function.

can be alleviated using the pseudo-likelihood or conditional independence in layer linking [10]. The original HGM in Fig. 6.3(b) can be approximated as shown in Fig. 6.6. For simplicity, the object layer and the part layer are represented as a single node. Mathematically, Eq. 6.2 can be rewritten as Eq. 6.3. Since each layer is conditioned on neighboring layers the global partition function is not required.

$$p(x^S, \mathbf{x}^O, \mathbf{x}^P | \mathbf{y}) \approx p(x^S | \mathbf{x}^O, \mathbf{y}) p(\mathbf{x}^O | x^S, \mathbf{x}^P) p(\mathbf{x}^P | \mathbf{x}^O, \mathbf{y}) \qquad (6.3)$$

The second problem for variable structure can be formulated by deriving each conditional probability in Eq. 6.3. If we condition based on object nodes $(\mathbf{x}_t^O)$ and observation nodes $(\mathbf{y})$, the likelihood of part layer $(\mathbf{x}_t^P)$ can be expressed as Eq. 6.4 by considering the incoming messages. $p(\mathbf{x}_t^P | y)$ can be regarded as a bottom-up message, and $p(\mathbf{x}_t^P | \mathbf{x}_t^O)$ can be regarded as a top-down message and α is the normalization factor. Fig. 6.6(c) is the graphical representation of Eq. 6.4.

$$p(\mathbf{x}_t^P | \mathbf{x}_t^O, y) = \alpha p(\mathbf{x}_t^P | y) p(\mathbf{x}_t^P | \mathbf{x}_t^O) \qquad (6.4)$$

By Bayes' rule, eEq. 6.4 becomes

$$p(\mathbf{x}_t^P | \mathbf{x}_t^O, y) = \alpha [p(y | \mathbf{x}_t^P) p(\mathbf{x}_t^P)] p(\mathbf{x}_t^P | \mathbf{x}_t^O) \qquad (6.5)$$

Although inference given an image is static, we consider a dynamic inference to reflect contextual influences in HGM. For example, if a part label is influenced by measurement then part-part context, and object-part context follow sequentially. Therefore, the prior $p(\mathbf{x}_t^P)$ can be predicted from the previous state $(p(\mathbf{x}_{t-1}^P))$ (see Fig. 6.7(c) and Eq. 6.6),

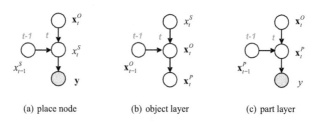

(a) place node   (b) object layer   (c) part layer

Figure 6.7. Modified pseudo-likelihood using Bayes' rule and previous prior states to reflect sequential contextual influence.

$$p(\mathbf{x}_t^P|\mathbf{x}_t^O, y) = \alpha p(y|\mathbf{x}_t^P)p(\mathbf{x}_t^P|\mathbf{x}_t^O) \int p(\mathbf{x}_t^P|\mathbf{x}_{t-1}^P)p(\mathbf{x}_{t-1}^P|\mathbf{x}_{t-1}^O, y)d\mathbf{x}_{t-1}^P \quad (6.6)$$

If we consider the trans-dimensional state jump (pruning unnecessary nodes) and pairwise bilateral interaction,

$$p(\mathbf{x}_t^P|\mathbf{x}_{t-1}^P) = \prod_i p(x_{it}^P|x_{i(t-1)}^P) \prod_{ij\in E_p} \psi(x_{it}^P, x_{jt}^P) \quad (6.7)$$

Summarizing Eq. 6.5-6.7 and considering graphical models in general (change the conditional likelihood to potential $\psi(x_i, x_j)$ indicating preferred pairs of values of directly linked variables x_i and x_j), Eq. 6.4 can be rewritten as Eq. 6.8. The current likelihood can be estimated from the measurement, object context, spatial context, and dynamic prediction prior.

$$
\begin{aligned}
p(\mathbf{x}_t^P|\mathbf{x}_t^O, y) = \alpha &\prod_i \phi(y_i, x_{it}^P)\psi(x_{it}^P, x_{it}^O) \prod_{ij\in E_p} \psi(x_{it}^P, x_{jt}^P) \\
&\times \int \prod_i p(x_{it}^P|x_{i(t-1)}^P)p(\mathbf{x}_{t-1}^P|\mathbf{x}_{t-1}^O, y)d\mathbf{x}_{t-1}^P
\end{aligned}
\quad (6.8)
$$

Likewise, the conditional likelihood of the object layer and place layer can be represented by Eq. 6.9 and 6.10 with corresponding simple graphical models, as shown in Fig. 6.7(b), 6.7(a). Note that all three layers (part, object, and place) consist of incoming contextual messages of a recursive nature. Note that the trans-dimensional state transition probabilities, such as $p(x_{it}^P|x_{i(t-1)}^P)$ and $p(x_{it}^O|x_{i(t-1)}^O)$, can treat the variable graph structures by recursive inference. Consequently, the second problem of handling variable structure can be solved.

$$p(\mathbf{x}_t^O | x_t^S, \mathbf{x}_t^P) = \alpha \prod_i \psi(\mathbf{x}_t^P, x_{it}^O) \psi(x_{it}^O, x_t^S) \prod_{ij \in E_O} \psi(x_{it}^O, x_{jt}^O)$$
$$\times \int \prod_i p(x_{it}^O | x_{i(t-1)}^O) p(\mathbf{x}_{t-1}^O | x_{t-1}^S, \mathbf{x}_{t-1}^P) d\mathbf{x}_{t-1}^O \qquad (6.9)$$

$$p(x_t^S | \mathbf{x}_t^O, \mathbf{y}) = \alpha p(\mathbf{y} | x_t^S) p(x_t^S | \mathbf{x}_t^O)$$
$$\times \int p(x_t^S | x_{t-1}^S) p(x_{t-1}^S | \mathbf{x}_{t-1}^O, \mathbf{y}) dx_{t-1}^S \qquad (6.10)$$

The place and object recognition problem in such graphical models is learning the graphical model of compatibilities and inferring the graph structure and marginal distributions from an image. We present a computationally feasible (approximate) method of learning and inference in next sections.

6.3.2   Generalized robust invariant feature

Like other chapters, we also utilize G-RIF as a basic feature detector since it shows better performance that SIFT empirically [35].

6.4   Piecewise Learning of HGM

6.4.1   Background of piecewise learning

Ideally, we would like to learn the model parameters by maximizing the conditional likelihoods in Eq. 6.8-6.10 from the given training images. This can be solved by gradient ascent requiring the evaluation of marginals of the hidden variables [29,40, 100]. This is computationally intractable due to the complexity of partition function. Instead of such conventional learning, we use piecewise learning, introduced by Sutton and McCallum [98]. Piecewise training involves dividing the undirected graphical model into pieces, each of which is trained independently. This is reported to be a successful heuristic for training large graphical models.

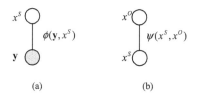

(a)   (b)

Figure 6.8. Types of piecewise learning in place node: (a) evidence of place label, (b) compatibility between place label and objects.

To train such large models efficiently, the basic concept of piecewise learning is to divide the full model into pieces that are trained independently, combining the learned weights from each piece of inference afterwards. According to Sutton and McCallum's proof and experiments on language data sets [98], the piecewise training method provides an upper bound on log partition functions, produces greater accuracy than pseudo-likelihood and performs comparably to global training using belief propagation. In computer vision, Freeman et al. [24] trained a graph model using piecewise training of compatibility functions for a super resolution application.

6.4.2   Piecewise learning in place node

For piecewise training related to place nodes, first we divide the graph as shown in Fig. 6.7(a) into two separate subgraphs of measurement and message from objects, as shown in Fig. 6.8.

Evidence of place label $(\phi(\mathbf{y}, x^S))$

We basically represent a place using histograms of learned visual words. First, we extract all local features from the labeled training set of images. Through a learning phase, in which an entropy-based MDL criterion is used, we obtain an optimal set of visual words (feature clusters) and class conditional distribution of visual words for inference [36]. Fig. 6.9 summarizes the steps for learning optimal visual words. There is only one parameter (ϵ) that controls the size of the visual words. Physically, ϵ is a similarity threshold used to measure descriptor distance in feature space. Through an iterative learning process, we can obtain the optimal visual words in terms of the entropy-based MDL criterion. If a novel object is presented, categorization is conducted using the detected features and learned visual words.

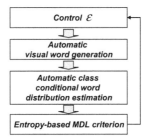

Figure 6.9. Visual word selection procedure for evidence of place label.

Kim and Kweon [36] introduce an entropy-based MDL criterion for simultaneous classification and visual word learning. The original MDL criterion is not
suitable since we have to find universal visual words for all places and sufficient
classification accuracy [105]. If the classification is discriminative, then the entropy of the class a posteriori probability should be low. Therefore, we propose an
entropy-based MDL criterion for simultaneous classification and visual word learning by combining MDL with the entropy of the class a posteriori probability, where
I denotes training images belonging to only one category. $\mathbf{V} = \{v_i\}$ denotes visual
words. N is the size of the training samples, and $\zeta(\mathbf{V})$ is the parameter size for
the visual words. Each visual word has parameters such as $\hat{\theta}_i = \{\mu_i, \sigma_i^2\}$ (mean,
variance). This MDL criterion is only useful to the class-specific learning of the
visual word (Low distortion with minimal complexity). Let $\mathbf{L} = \{(I_i, c_i)\}_{i=1}^{N}$, a set
of labeled training images where $c_i \in \{1, 2, \cdots, C\}$ is the class label. Then, the
entropy-based MDL criterion is defined as Eq. 6.11. λ represents the weight of
complexity term.

$$\hat{\mathbf{V}} = \arg\min_{\epsilon} \left\{ \sum_{i}^{N} H(c|I_i, \hat{\Theta}_{(\mathbf{V})}(\epsilon)) + \lambda \cdot \zeta(\mathbf{V}) \frac{\log(N)}{2} \right\} \qquad (6.11)$$

where entropy H is defined as

$$H(c|I_i, \hat{\Theta}_{(\mathbf{V})}) = -\sum_{c_i=1}^{C} p(c|I_i, \hat{\Theta}_{(\mathbf{V})}) \log_2 p(c|I_i, \hat{\Theta}_{(\mathbf{V})}) \qquad (6.12)$$

In Eq. 6.11, the first term represents the overall entropy for the training image
set. The lower the entropy, the better the guaranteed classification accuracy; the

second term acts as a penalty on learning. If the size of the visual words increases, then the model requires more parameters. Therefore, if we minimize Eq. 6.11, we can find the optimal set of visual words for successful classification with moderate model complexity. Note that we require only one parameter ϵ for minimization as ϵ controls the size of the visual words automatically. Details are given in the next section.

Since ϵ is the distance threshold between the normalized features, we can obtain initial clusters with an automatic sizing. We can obtain refined cluster parameters $(\hat{\Theta}_{\mathbf{V}}) = \{\theta_{(i)}\}_{i=1}^{V})$ with k-mean clustering. After ϵ-NN-based visual word generations, we have estimated the class-conditional visual word distribution for the entropy calculation. The Laplacian smoothing-based estimation is defined by Eq. 6.13 [13].

$$p(v_t|c_j) = \frac{1 + \sum_{(I_i \in c_j)} N(t,i)}{V + \sum_{s=1}^{V} \sum_{I_i \in c_j} N(s,i)} \qquad (6.13)$$

where $N(t,i)$ represents the number of occurrences of the visual word (v_t) in the training image (I_i), and V represents the size of the visual words. The physical meaning of this equation is the empirical likelihood of visual words for a given class.

Finally, we can calculate the posterior $p(c|I_i, \hat{\Theta}_{(\mathbf{V})})$, which is used for the entropy calculation in Eq. 6.11. Using Bayes rule and uniform distribution of class lables, this can be stated as Eq. 6.14 where the image is approximated with a set of local features $(I_i \approx \{y\}_i)$.

$$p(c|I_i, \hat{\Theta}_{(\mathbf{V})}) = \alpha p(I_i|c_i, \hat{\Theta}_{(\mathbf{V})})p(c_i) \approx \alpha p(\{y\}_i|c_i, \hat{\Theta}_{(\mathbf{V})}) \qquad (6.14)$$

Using the naive Bayes' method (assuming independent features),

$$p(\{y\}_i|c_i, \hat{\Theta}_{(\mathbf{V})}) = \prod_j \sum_{t=1}^{V} p(y_j|v_t)p(v_t|c_i) \qquad (6.15)$$

where $p(y_j|v_t) = exp\left\{-\|y_j - \mu_t\|^2/2\sigma_t^2\right\}$

Form the calculations defined by Eq. 6.12-6.15, we can evaluate Eq. 6.11. We learn the optimal set of visual words by changing and evaluating Eq. 6.11 iteratively. We found that the optimal is 0.4 for scene visual words. The evidence of place, , is the same as Eq. 6.15.

Compatibility of place label and object label $(\psi(x^S, x^O))$

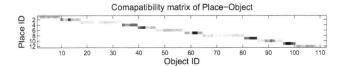

Figure 6.10. Compatibility table of places and objects.

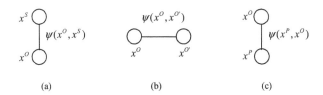

Figure 6.11. Types of piecewise learning in object layer: (a) compatibility between object and place, (b) compatibility between objects, (c) compatibility between part and object.

The conditional likelihood of place labels for given objects can be estimated by counting the number of object appearances at places. We add a Dirichlet smoothing prior to the count matrix so that we do not assign zero likelihood that does not appear in the training data. Fig. 6.10 shows the probability look-up table for place-object, given labeled images (12 places with 112 objects). Black represents high strong compatibility between place and object.

6.4.3   Piecewise learning in the object layer

The graphical model of the object layer shown in Fig. 6.7(b) can be subdivided into three subgraphs as shown in Fig. 6.11. It consists of object likelihood given place, object-object compatibility, and part likelihood for the given object.

Compatibility between object and place ($\psi(x^O, x^S)$)

Compatibility between object and place is obtained directly from the probability look-up table of place and object shown in Fig. 6.10. From the table, we can predict the object likelihood given a specific place, by renormalizing each row in the table.

Compatibility between objects ($\psi(x^O, x^{O'})$)

We estimate the compatibility matrix empirically by counting co-occurrence of

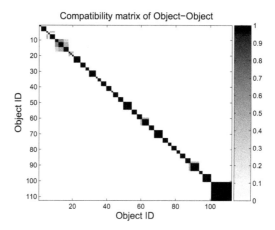

Figure 6.12. Object-object compatibility matrix obtained from labeled training set.

objects. Initially, we count co-occurrence of objects for each training image. We also add a Dirichlet smoothing prior to the count matrix so that we do not assign zero likelihood that does not appear in the training data. Then the matrix inner product is normalized using diagonal values so that the self-compatibility is unity. Fig. 6.12 shows the object-object compatibility matrix. Objects that co-occurred spatially are clustered diagonally. Object-object compatibility is used in Eq. 6.9 for spatial interaction of objects.

Compatibility of part and object ($\psi(x^P, x^O)$)

Physically, $\psi(x^P, x^O)$ represents a top-down message from object model to parts in terms of the image position of the parts. To derive this likelihood, we utilize a robust object representation scheme introduced in scalable object representation chapter to handle the object recognition issues.

Given such part-based object representation and learning, we can estimate the likelihood of $\psi(x^P, x^O)$ as follows. At inference given arbitrary test image, x^O contains both object label and pose (θ_O) of object frame and x^P contains observed part pose. Therefore, we predict the part positions ($\chi_M = \theta_O(\chi(p_i))$) due to the property of the common-frame constellation model, where p_i represents a part in a CFCM and $\chi()$ represents the pose term of ($\cdot$). The positions of the model parts depend on the estimated object pose (θ_O). Finally, $\psi(x^P, x^O)$ is defined by Eq.

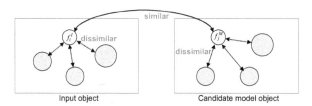

Figure 6.13. The concept of saliency-based visual distance measure for stable matching.

6.16, where σ_χ is standard deviation of part position learned during the CFCM construction.

$$\psi(x^P, x^O) = \exp\left(-\frac{\|\chi(x^P) - \theta_O(\chi(p_i))\|^2}{2\sigma_\chi^2}\right) \qquad (6.16)$$

In this likelihood estimation, it is essential to estimate the object pose θ_O accurately. For a given input image and a candidate object model, conventional methods usually find corresponding points using Euclidean feature distance. However, if there are similar parts within an object, we cannot obtain the correct pose by a similarity transform between the input and the model. We propose a novel visual distance measure based on the concept of saliency and extending Lowe's distance ratio, which considers only the first and second nearest neighbors [54]. If an image feature (f_i^I) and a corresponding model feature (f_j^M) are sufficiently similar and f_i^I is quite different from the rest of the input features and f_j^M is also quite different from the rest of the model features, then the matching is very reliable (see Fig. 6.13). Based on this concept, we propose a novel saliency-based distance measure (D_P) defined by Eq. 6.17, D_E where represents the Euclidean distance and $C(f_i^I)$ represents contrast or saliency of f_i^I within the input feature set, which is defined as the relative distance ratio between the nearest and second nearest neighbors. As the D_P is smaller, we produce more stable matches.

$$D_P(f_i^I, f_j^M) = D_E(f_i^I, f_j^M) \cdot (C(f_i^I) + C(f_j^M))$$

$$where\ C(f_i^I) = \frac{D_E^{1stNN}(f_i^I, \{f^I\})}{D_E^{2ndNN}(f_i^I, \{f^I\})}, C(f_j^M) = \frac{D_E^{1stNN}(f_j^M, \{f^M\})}{D_E^{2ndNN}(f_j^M, \{f^I\})} \qquad (6.17)$$

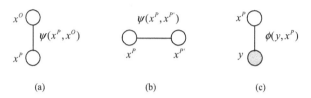

Figure 6.14. Types of piecewise learning in part layer: (a) compatibility between part and object, (b) compatibility between parts, (c) evidence of part.

6.4.4 Piecewise learning in part layer

The graphical model of the part layer is shown in Fig. 6.7(c), which can be subdivided into three subgraphs as shown in Fig. 6.14. It consists of part likelihood given object, part-part compatibility, and feature likelihood for a given part.

Compatibility between part and object ($\psi(x^P, x^O)$): The message to part layer from object layer shown in Fig. 6.11(a) is the same as Eq. 6.16.

Part-part compatibility ($\psi(x^P, x^{P'})$): We assume a pairwise clique as shown in Fig. 6.11(b). In MRF, the compatibility is defined as the energy of label smoothness [49]. In DRF, it is defined as the interaction energy of the part position, including label smoothness, and parameters are learned by gradient ascent [40]. Instead of such modeling, we apply the gestalt law of proximity and similarity introduced in Chapter 4 because parts within an object are geometrically very close and have the same object labels, as shown in Fig. 6.15. We model such a property as part-part compatibility, as defined by Eq. 6.18, where the first term reflects the weight for spatial distance, the second term reflects weight for same object labeling. If two parts have different labels, we assign very small weight (ζ). σ_D is defined as the standard deviation of distances between part pair in the given training set. The labeling similarity weight is calculated using $\phi(y, x^P)$, defined in the following section. $ID(\cdot)$ represents the part label of part ($\cdot$).

$$
\psi(x_T^P, x_S^P | ID(x_T^P) = i, ID(x_S^P) = i)
$$
$$
\propto \exp\left(\frac{\|\chi(x_T^P) - \chi(x_S^P)\|^2}{2\sigma_D^2}\right) \cdot \frac{w(ID(x_T^P) = i, ID(x_S^P) = i)}{\sum_j w(ID(x_T^P) = j, ID(x_S^P) = j)} \qquad (6.18)
$$

where $w(ID(x_T^P) = i, ID((x_S^P) = i) = \phi(y_T, ID(x_T^P) = i)\phi(y_S, ID(x_S^P) = i)$

Evidence of part ($\phi(y, x^P)$): The likelihood of an observed feature's appearance in a given part node is calculated from Eq. 6.19. Since the part node is given,

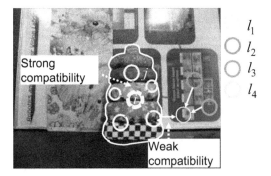

Figure 6.15. The concept of part-part compatibility is based on the proximity and similarity of the gestalt principle. If a part pair has small distance and the same part labeling, the part-part compatibility is strong.

we know which appearance library is used from the memorized linking ($\psi(x^P, A)$), A where denotes the appearance library; we assume noisy measurement for the likelihood, A and σ_A are learned during ϵ-nearest neighbor clustering.

$$\phi(y, x^P) = G(y|A)\psi(x^P, A), where G(y|A) = \exp\left(-\frac{\|A(y) - A\|^2}{2\sigma_A^2}\right) \quad (6.19)$$

6.5   Approximate Inference for Online Graph Construction

6.5.1   Multi-modal sequential Monte Carlo (MM-SMC)

The inference in this chapter is a very difficult problem because we have to estimate both the graph structure and a posteriori probability distribution, as different images have different numbers of objects and parts. If the graph structure is fixed and a posteriori probability distribution is available, Eq. 6.1, then maximum a posteriori probability (MAP) can give an exact inference. In practice, it is not trivial to estimate a posteriori probability, even for a fixed graph, due to the loopy graph structure. Most researchers use approximate methods to estimate a posteriori probability. One is to use variational inference methods [95], which factorize an a posteriori probability distribution into a simpler distribution. Another method is to use Monte Carlo (sampling) approximation, such as Markov Chain Monte Carlo [27], where a Markov chain is constructed that has the desired a posteriori probabilities as the limit distributions. A third approach is to use loopy belief propagation [119] to approximate a posteriori probability marginals for loopy graphs.

If the graph structure varies from image to image, it cannot be solved by directly applying these approximate inference methods. Recently, extended versions of Monte Carlo methods have been proposed to handle graph variability and a posteriori probability approximation. They are the Reversible Jump MCMC (RJ-MCMC) sampler [103] and the Trans-dimensional SMC (TD-SMC) sampler [111]. The RJ-MCMC is theoretically more rigorous (no normalization), but it requires a very long time for the inference of high dimensional states. TD-SMC is similar to RJ-MCMC but is more flexible because TD-SMC has no constraint on reversibility. The complexity is the same as for RJ-MCMC.

In this chapter, we use several approximation schemes for the simultaneous inference of graph structure and marginal. First, we approximate the original a posteriori probability (Eq. 6.2) as three conditional likelihoods or pseudo-likelihoods using Eq. 6.3. This simplifies the global partition function into subpartition functions. Second, we further approximate the joint pseudo-likelihood (conditional likelihoods) in Eq. 6.8-6.10 by loopy belief propagation (LBP) to estimate a posteriori probability marginals [119]. BP provides an approximate solution for the loopy graphical model and empirically shows successful performance. Through the LBP, we obtain an approximate solution by maximizing the a posteriori probability marginal (MPM). In the original BP, belief (b_i) at node i is formulated as Eq. 6.20 (see also Fig. 6.16). In the message calculation, we approximate $\phi_j(x_j) \prod_{k \in N(j) \backslash i} m_{kj}(x_j)$ by $b_j^{t-1}(x_j)$ estimated in the previous time step and we use max-product instead of sum-product for accurate estimation.

$$
\begin{aligned}
b_i(x_i) &= k\phi_i(x_i) \prod_{j \in N(i)} m_{ji}(x_i) \\
m_{ji}(x_i) &\leftarrow \sum_j \psi_{ij}(x_i, x_j)\phi_j(x_j) \prod_{k \in N(j) \backslash i} m_{kj}(x_j)
\end{aligned}
\tag{6.20}
$$

Third, we finally approximate the a posteriori probability marginal into multimodal sequential Monte Carlo. The multi-modal scheme can handle the uncertainty of the graph structure and marginal distribution simultaneously [109]. Since the

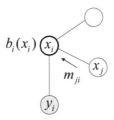

Figure 6.16. Basic concept of belief propagation.

marginal in each node cannot be represented parametrically due to non-Gaussian measurement and occlusions, we use sequential Monte Carlo (SMC). In the Eq. 6.8-6.10, this has a recursive filtering nature so SMC is suitable for our problem solution. Therefore we call our approximate inference multi-modal sequential Monte Carlo (MM-SMC). Physically, the MM-SMC does not conclude immediately, but to allow multiple probable hypotheses (particles). These particles stay alive until longer feedback (scene context) is influenced. A node has multiple particles which contain hypotheses with weights. Each weight is updated by belief propagation (BP) [46].

6.5.2   Hypothesis and prune for structure estimation in MM-SMC

We propose a hypothesize and prune method rather than a Markov Chain to fully utilize the contextual influence during inference. As described above, a Markov Chain, such as RJ-MCMC, is very slow and does not fully utilize the spatial and hierarchical context. In the hypothesize and prune method, we initialize a graph structure which contains the true graph structure using the bottom-up method (see Fig. 6.17(a)). Then the distribution of each graph node is represented by a set of samples (Monte Carlo) (see Fig. 6.17(b)). Each sample weight is updated using the BP in MM-SMC. During this process, the spatial context and hierarchical context are activated. Contextually consistent graph nodes receive strong weights and contextually inconsistent graph nodes receive very weak weights (see Fig. 6.17(c)). After each contextual influence, we conduct the on-the-fly structure pruning to remove the wrong graph nodes (see Fig. 6.17(d)). Note that several nodes can be removed simultaneously instead of one by one as in RJ-MCMC. This is the reason hypothesis pruning is computationally more efficient than RJ-MCMC.

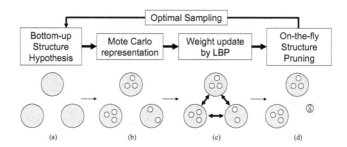

Figure 6.17. Simultaneous estimation of graph structure and marginal distribution using the MM-SMC method: (a) hypothesis of graph structure, (b) sample generation for each marginal distribution, (c) weight calculation using LBP in MM-SMC, (d) on-the-fly structure pruning.

6.5.3   Implementation details of MM-SMC

Particle design: A particle in a place node represents a probable place label. Particles in an object node represent possible multiview CFCMs, which contain object IDs with poses (similarity transform parameters: position, scale, orientation). Particles in a part node represent probable object IDs where they belong with observed poses (position, scale, orientation). All the pose parameters for objects and parts are estimated deterministically from image structure (object pose: similarity transform between model and image, part pose: image feature) to reduce dimensionality.

Representation of MM-SMC: We formally represent the MM-SMC, which is a nonparametric form of multi-modal. Assume the number of multi-modals is M, note that initially we do not know the true number of objects, which must be estimated through the hypothesis and pruning strategy (so the number should be larger than the true number of objects). The following scheme is almost the same as for the part layer. An object node collects messages from the part layer ($\mathbf{x}^P$), scene layer (x^S), neighboring objects and messages from the previous stage. Let $O = \{N, M, \Pi, X, W, C\}$ denote the particle representation of the object layer, with N the number of total view-tuned particles (CFCM), M the number of mixture components, $\Pi = \{\pi_m\}_{m=1}^{M}$ the mixture component (object) weights, $X = \{x_{(i)}^o\}_{i=1}^{N}$ the particles, $W = \{w_{(i)}\}_{i=1}^{N}$ the particle weights, and $C = \{c_{(i)}\}_{i=1}^{N}$ the object indicators, i.e. $c_{(i)} \in \{1, 2, \cdots, M\}$. If particle i belongs to mixture component m, then

$c_{(i)} = m$. $O_m = \{i \in \{1, \cdots, N\} : c_{(i)} = m\}$ is the set of indices of the particles belonging to the m-mixture components (objects). When M is the number of objects (or graph nodes) in a scene, each mixture component has view-tuned particles with different poses. The approximate multi-modal particles can be represented by Eq. 6.21.

$$b(\mathbf{x}^O) = \sum_{m=1}^{M} \pi_m b(x_m^O) \approx \sum_{m=1}^{M} \pi_m \sum_{i \in O_m} w_{(i)} \delta x_{(i)}^o \qquad (6.21)$$

Bottom-up graph hypothesis (node birth) and particle generation: We generate particles using importance sampling, especially the data-driven proposal function, for fast and accurate convergence [109]. In the part layer, the initial part graph is hypothesized by part detector in G RIF [35]. G-RIF can provide corner parts and convex parts that are complementary. A particle in a part node has an object label with observed part pose. Particles (different object labels with the same part pose) in each part node are generated automatically using the visual appearance library because visual codebooks contain all the links to the object models (CFCMs). We use ϵ-nearest neighbor search for the appearance library (ϵ=0.2) and the number of part particles in each part node is 5-10. In the object layer, the data-driven proposal function (q) for object particles is defined by Eq. 6.22. The Hough transform of the part particles generates CFCMs in pose space, as shown in Fig. 6.18(b) and 6.18(c). We usually have 2 6 object particles per node. Note that these particles are very useful and can provide quick convergence because the particles are generated in a data-driven way.

$$q(\mathbf{x}_t^O | x_t^S, \mathbf{x}_x^P) \sim Hough(\mathbf{x}_t^O | \mathbf{x}t^P) \qquad (6.22)$$

Weight calculation by LBP in MM-SMC: Importance weight in each node is updated by combining incoming messages, as shown in Fig. 6.19. We calculate each message using Eq. 6.23 and weights are updated using Eq. 6.24, represents belief or weight of the considering particles.

$$
\begin{aligned}
M_1(x_{(i)}^o) &= \max_k \{b(\{x_{(k)}^P\})\psi(\{x_{(k)}^p\}, x_{(i)}^o)\} \\
M_2(x_{(i)}^o) &= \max_k \{b(x_{(k)}^s)\psi(x_{(k)}^s, x_{(i)}^o)\} \\
M_3(x_{(i)}^o) &= \prod_{l \in \Gamma(m)} M_{lm}(x_{(i)}^o) \\
where M_{lm}(x_{(i)}^o) &= \max_k \{b(x_{(k)}^o)\psi(x_{(k)}^o, x_{(i)}^o)\}
\end{aligned}
\qquad (6.23)
$$

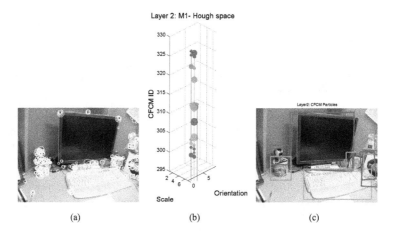

Figure 6.18. Data-driven initialization of graph structure in part layer using the G-RIF part detector (a), and object layer using Hough transform in pose space (b) and in image space (c).

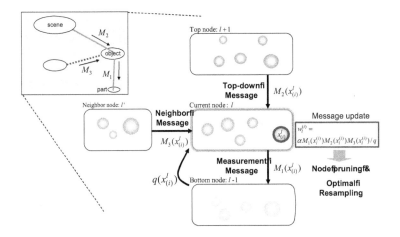

Figure 6.19. The diagram of importance sampling-based LBP in MM-SMC at the object layer.

$$w_{(i)}^{new} = \frac{\tilde{w}_i}{\sum_{j \in O_m} \tilde{w}_j}, \ \tilde{w}_i = M_1(x_{(i)}^o) \cdot M_2(x_{(i)}^o) \cdot M_3(x_{(i)}^o)/q(x_{(i)}^o)$$

$$\pi_m^{new} \approx \frac{\pi_m \tilde{w}_m}{\sum_{n=1}^M \pi_n \tilde{w}_n}, \ \tilde{w}_m = \sum_{i \in O_m} \tilde{w}_{(i)} \tag{6.24}$$

Graph pruning (node death) and resampling: After calculating the importance weight using MM-SMC, we conduct two steps of sample selection for graph structure estimation and marginal distribution. First, the multi-modal components with very small weights ($\pi_m \le \rho$) are removed by contextual influences. Nodes that are contextually inconsistent receive little weight. Second, we conduct optimal sampling from the surviving multimodal densities using optimal resampling [20]. If a sample is reselected, then $p(x_{i,t}^O | x_{i,t-1}^O)$ equals unity during MM-SMC. Note that our inference is static not temporal tracking where $p(x_{i,t}^O | x_{i,t-1}^O)$ is the motion a priori probability. Competing particles survive until longer contextual feedback messages are activated. Through the data-driven graph structure hypothesis and pruning by contextual influences, we can produce the true inference.

Overall inference algorithm: Our inference system conducts the hypothesis pruning in place layer, object layer, and part layer in parallel to produce the true graph structure and maximum a posteriori probability marginal (MPM) solution. The hypothesis is governed by bottom-up information and pruning is governed by contextual influences.

6.6 Experimental Results

6.6.1 Validation of bidirectional reinforcement property

First, we tested the bidirectional place and object recognition method on ambiguous examples. If we used object information only, as shown in Fig. 6.20(a), we fail to discriminate both objects since local features are almost identical. However, if we used the bidirectional interaction method, we could discriminate between those objects simultaneously, as shown in Fig. 6.20(b).

In the second experiment, we prepared place images taken in front of an elevator. Given a test elevator scene, as shown in Fig. 6.21(a), measurement only provided incorrect place recognition. However, if the same test scene was processed using bidirectional place-object recognition we produced the correct recognition, as shown in Fig. 6.21(b). In the diagram, the center graph shows the measurement message, the right graph shows the message from the objects, and the left graph represents the combined message for place recognition.

6.6.2 Large scale experiment for building guidance

We validated the proposed scene interpretation method in terms of false alarm reduction. 620 object images (112 objects) were acquired in 12 topological places, such as offices, corridors, etc. (a total of 228 images for test of 648×480 pixels), as shown in Fig. 6.22. Note that the images are completely general and obtained in uncontrolled environments. We trained using this labeled training set and the piecewise learning methods described in Section 4. We used 114 test images that were not used in training; the test set contains 645 object images. After training, the size of the appearance library was reduced by 33.3% from 72,083 to 48,063 (ϵ=0.2). After the shared-feature-based view clustering, the CFCM size was reduced from 5.5 CFCMs/object to 2.4 CFCMs/object ($T2$ =15).

False alarm reduction by saliency-based pose estimation

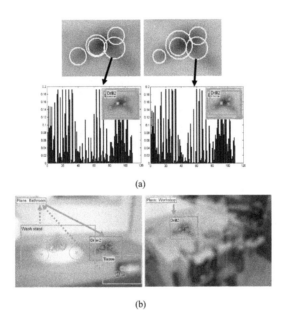

(a)

(b)

Figure 6.20. Toy example for collaborative property (a) Object recognition using only object-related features [4] (b) Application of bidirectional interaction between place and objects. Dotted arrows represent objects to place message to disambiguate place. Solid arrows represent place to object message to disambiguate blurred objects (drier, drill).

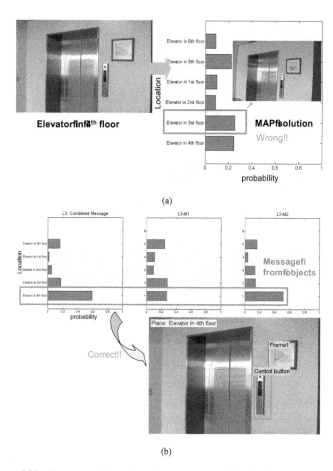

Figure 6.21. Place recognition using only measurement (a) and both measurement and message from objects (b).

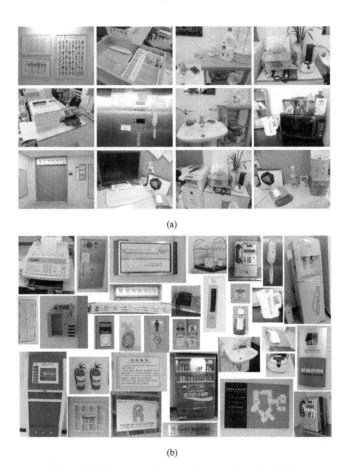

(a)

(b)

Figure 6.22. Examples of scenes (a), and related objects (b).

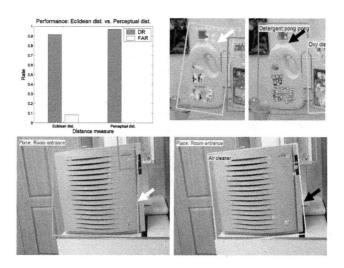

Figure 6.23. (Left top) Performance evaluation using Euclidean distance and the proposed perceptual distance. Pairs of images show the pose estimation using corresponding distance measures. White arrows indicate incorrect pose estimation using Euclidean distance and black arrows indicate pose estimation using our method.

We evaluated the influence of the proposed salience-based perceptual distance on the pose calculation between model objects and images. As a performance evaluation measure, we used a detection rate (DR: correct name + correct pose) and false alarm rate (FAR: incorrect name or incorrect pose). Fig. 6.23 shows the power of the proposed perceptual distance to reduce false alarms generated by incorrect poses. In this test, we used a complete inference model incorporating scene, object, part context; the white arrow indicates pose estimation using Euclidean distance and the yellow arrow indicates the result of using proposed distance. Note that repeated parts can cause incorrect pose but our method can estimate object pose using salient parts (note the air cleaner at the bottom of Fig. 6.23).

Overall performance evaluation

In the second evaluation, we validated the power for false alarm removal using the proposed context-based inference scheme. The previous results show only a partial effect of scene context. There are four contexts, part context ($L1M3$: neighboring message), object context of top-down ($L1M2$), object context from neighboring objects ($L2M3$), and scene context ($L2M2, L3M2$). The basic scene interpretation block is $C1$ which consists of part evidence ($L1M1$) and bottom-up message for objects ($L2M1$). We call this the baseline method, which has almost no context. The baseline method is equivalent to SIFT with Hough transform with a very small bin threshold [54]. In first test, we evaluated the proposed system by adding four kinds of messages one by one. Fig. 6.24(a) shows the DR and FAR results. The basic scene interpretation block shows 26% false alarms for 645 test object images. However, by adding higher contexts to this basic block, the FAR is reduced to 0.15% (only 1 false alarm) for full contexts. In the next test, we checked what contributions were provided by each context to the false alarm reduction problem. We conducted tests similar to previous tests but we combined each context to the basic inference block ($C1$) separately. Fig. 6.24(b) summarizes the evaluation results. The contribution is sorted as part context ¿ object context of top-down ¿ scene context ¿ object context from neighboring objects. Fig. 6.25 shows real scene interpretation results without scene context and with scene context. The proposed system can remove false alarms effectively and provides topological place information.

Fig. 6.26 shows partial examples of scene interpretation in an indoor environment. The proposed method can provide object and place information simultaneously. As can be seen, the related information is extracted stably in various places. In this test, all the place information is recognized correctly. Consequently, it can be used in vision-based guidance systems for visitors or mobile robot applications. The average inference time is 50 sec at AMD 4800+ machine in a MATLAB environment. Most inference time is consumed in database search (over 35 sec) and core inference is very fast.

6.7   Summary

This chapter proposed a collaborative place, object, and part recognition system that uses spatial and hierarchical contexts in an HGM. Partial contexts introduced in previous chapters are incorporated. Our HGM has a three-layered structure, composed of place, object, and part layers. The object layer and part layer have a pose-encoded-MRF model to reflect the spatial context. These layers are indirectly linked to reflect the hierarchical context. We train the graph using piecewise learning for tractability. We incorporate a robust and scalable object representation scheme to

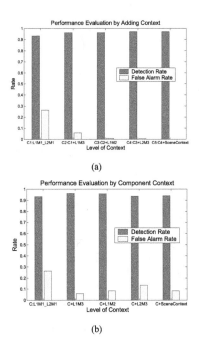

(a)

(b)

Figure 6.24. Performance evaluation according to various contexts: (a) by adding higher contexts, (b) the effect of individual context. Notations are indicated in the text.

Figure 6.25. Scene interpretation results: (a) without scene context, (b) with scene context

handle visual variations. For successful inference (simultaneous graph structure estimation with marginal), we adopt a pseudo-likelihood strategy and MM-SMC. Because of the data-driven hypothesis generation and contextual influence-based pruning, we can produce accurate inference. In addition, we propose a novel visual distance measure based on the concept of saliency. We obtain correct point correspondences to estimate object poses. We validated the false alarm reduction methods by applying the method to real scene data. The proposed system showed only one false alarm for 645 test objects. Next chapter introduces how to combine the static context and temporal context in video. In Chapter 8 and 9, we will investigate upgrading the current scene interpretation at the identification level to a categorization level based on the HGM.

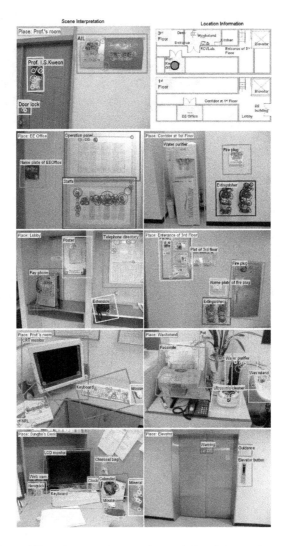

Figure 6.26. Various scene interpretation results in an indoor environment

Chapter VII

Static and Temporal Context: Video Intepretation

In this chapter, we present a practical place and object recognition method for guiding visitors in building environments by modeling static and temporal context. Due to motion blur or camera noise, places or objects can be ambiguous. The first key contribution of this work is the modeling of bidirectional interaction between places and objects for simultaneous reinforcement. The second key contribution is the unification of visual context, including scene context, object context, and temporal context. The last key contribution is a practical demonstration of the proposed system for visitors in a large scale building environment.

7.1   Visual ambiguity

Let us imagine that a visitor is looking around a complex building. He might need a guide to get place and object information. This can be realized by using a wearable computer and recent computer vision technology. A web camera on the head receives video data and a wearable computer processes the data to provide place and object recognition results to the person in the form of image and sound in the head mount display (HMD). Visitors with such human computer interaction (HCI) devices can get information on objects and find specific places quickly.

Although the general purpose of place and object categorization is not possible with current state-of-the-art vision technology, recognition or identification of places and objects in certain environments is realizable because of the development of robust local features [35,54] and strong classifiers like SVM and Adaboost [113]. However, there are several sources of ambiguities caused by motion blur, camera noise, and environmental similarity. Fig. 7.1 shows an example of place ambiguities caused by similar environments. Recently, Torralba et al. and Murphy et al. proposed context-based place and object recognition methods [67, 102]. In [67], Murphy et al. developed a tree structure-based scene, object recognition method

Figure 7.1. Ambiguity of places by similar environments (Which floor are we at?). This can be disambiguated by recognizing specific objects (pictures on the wall).

by incorporating gist and boosting information. In [102], Torralba et al. utilized gist information from the whole scene. This gist provides strong prior object label and positioning information. These approaches attempted to solve the ambiguity of objects using scene information.

However, no one has tried to disambiguate place label explicitly. Only Torralba incorporated temporal context, which was modeled as the Hidden Markov Model (HMM) [102]. In this chapter, we focus on the disambiguation method of simultaneous place and object recognition using bidirectional contextual interaction between places and objects [4]. The human visual system (HVS) can recognize place immediately using a low amount of spatial information. If the place is ambiguous, HVS can discriminate the place using object information in the scene (see the rectangle regions in Fig. 7.1). Motivated from this bidirectional interaction, we present a more robust place and object recognition method.

7.2   Place Recognition in Video

7.2.1   Graphical model-based formulation

Conventionally, place labels from video sequences can be estimated by the well-known Hidden Markov Model (HMM) [102]. We extend the HMM by incorporating the bidirectional context of objects. You can get a clearer concept of the extension through the graphical model, especially Bayesian Net, as shown in Fig. 7.2. A place node at time t is affected by three kinds of information: measurement message (likelihood) from an image, top-down message from objects, and temporal message from the previous state. Let $Q_t \in \{1, 2, \ldots, N_p\}$ represent place label at

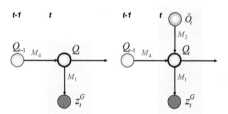

Figure 7.2. Conventional graphical model (left) and extended graphical model (right) for place recognition in video. Belief at the place node (center circle) gets information from image measurement (M_1), object information (M_2), and previous message (M_4).

t, z_t^G represent whole image features, $\vec{O}_t$ represent object label vector, and $T(q', q)$ represent the place transition matrix. The Bayesian formula for this graphical model can be represented by Eq. 7.1.

$$p(Q_t = q | z_{1:t}^G, \vec{O}_{1:t}) \propto p(z_{1:t}^G | Q_t = q) p(Q_t = q | \vec{O}_{1:t}) p(Q_t = q | z_{1:t-1}^G, \vec{O}_{1:t-1})$$
$$where\ p(Q_t = q | z_{1:t-1}^G, \vec{O}_{1:t-1}) = \sum_{q'} T(q', q) p(Q_{t-1} = q' | z_{1:t-1}^G, \vec{O}_{1:t-1})$$

$$(7.1)$$

$p(z_{1:t}^G | Q_t = q)$ represents bottom-up messages (measurement) from whole images (M_1), $p(Q_t = q | \vec{O}_t)$ represents top-down messages coming from object label (M_2), $p(Q_t = q | z_{t-1}^G, \vec{O}_{t-1})$ represents temporal messages from the previous state (M_4). M_2 is calculated by combining messages from related objects using the scene-object compatibility matrix. The important thing from Eq. 7.1 is how to utilize individual messages. Combining all the messages is not always a good idea in terms of performance and computational complexity. We can think of three kinds of situations: no temporal context is available (ex. initialization, kidnapped), static context (bottom-up, top-down) is useless due to blurring, and static and temporal context are available and necessary. Since we do not know such situations a priori, we propose a stochastic place estimation scheme as Eq. 7.2. γ is the probability of reinitialization (mode1) where temporal context is blocked. α is the probability of normal tracking (mode2) where static context is prevented, which reduces the computational load. Otherwise, mode 3 is activated for place estimation. For each frame, each mode is selected according to the selection probability. In this chapter, we set the parameters as $\gamma = 0.1, \alpha = 0.8$ by manual tuning.

$$Model1(\gamma) : p(Q_t = q|z_{1:t}^G, \vec{O}_{1:t}) \propto M_1 M_2$$
$$Mode2(\alpha) : p(Q_t = q|z_{1:t}^G, \vec{O}_{1:t}) \propto M_1 M_4 \qquad (7.2)$$
$$Mode3(1 - \gamma - \alpha) : p(Q_t = q|z_{1:t}^G, \vec{O}_{1:t}) \propto M_1 M_2 M_4$$

7.2.2   Modeling of measurement (M_1)

There are two kinds of place measurement methods depending on feature type. Torralba et al. proposed a very effective place measurement using a set of filter bank responses [102]. In this chapter, our place-object recognition system is based on local features proposed in Chapter 3 [35]. Generalized robust invariant feature (G-RIF) is a generalized version of SIFT [54] by decomposing a scene into convex parts and corner parts, which are described by localized histograms of edge, orientation, and hue. The G-RIF shows upgraded recognition performance by 20% than SIFT for COIL-100 [37]. For place classification, we utilize the bags of keypoints method proposed by Csurka et al. [13]. Although the SVM-based classification shows better performance than naive Bayes [113], we use a naive Bayes classifier to show the effect of context. The incoming messages from objects or previous places can compensate for the rather weak classifier.

7.2.3   Modeling of object message (M_2)

Direct computation of object messages from multiple objects is not easy. If we use the graphical model [32, 119], we can estimate approximate messages. Since the number (N_O) of objects and distribution of object nodes are given, the incoming message M_2 to a place node is expressed as Eq. 7.3.

$$M_2(Q_t = q) = p(Q_t = q|\vec{O}_t) = \prod_{i=1}^{N_O} p(Q_t = q|O_t^i)$$
$$where \; p(Q_t = q|O_t^i) \propto \max_k \{\psi(q, O_t^i(k))p(O_t^i(k))\} \qquad (7.3)$$

$O_t^i(k)$ is a hypothesis of multi-views for 3D object O_t^i where k is multi-view index. $p(O_t^i(k))$ is estimated probability of the object hypothesis. $\psi(q, O_t^i(k))$ is the compatibility matrix of the place label and object label. It is estimated by counting the co-occurrences from place-labeled object data. Eq. 7.3 is an approximated version of belief propagation. The max-product method is incorporated instead of the sum-product for the increased estimation accuracy. Physically, the individual maximum message ($p(Q_t = q|O_t^i)$) from each object is combined to generate the objects to scene message (M_2).

7.2.4 Modeling of temporal message (M_4)

The computation of temporal message from previous place to probable current place is defined as Eq. 7.4. It is the same equation used in the HMM [102]. The place transition matrix $T(q', q)$ is learned by counting frequencies in the physical path. This term prevents quantum jumps during place recognition.

$$M_4(Q_t = q) = p(Q_t = q|z_{t-1}^G, \vec{O}_{t-1}) = \sum_{q'} T(q', q) p(Q_{t-1} = q'|z_{t-1}^G, \vec{O}_{t-1})$$

$$(7.4)$$

7.3 Object Recognition in Video

7.3.1 Graphical model-based formulation

In general, an object node in video can get information from measurement (M_1), message from scene (M_2), and information from the previous state (M_4) as shown in Fig. 7.3 (c), which is combined version of (a) and (b). Murphy et al. proposed a mathematical framework for combining M_1 and M_2 with a tree structured graphical model as in Fig. 7.3 (a) [67]. Vermaak et al. proposed a utilization method for combining M_1 and M_4 to track multiple objects as Fig. 7.3 (b) [109].

To our knowledge, this is the first attempt to unify these messages within a graphical model. We assume independent objects for simple derivation (see [34] for interaction). This is reasonable since objects are conditioned on a scene. Therefore, we consider only one object for simple mathematical formulation. Let $X_t = (O_t, \theta_t)$ represent a hybrid state composed of an object label and its pose at t. The pose is the similarity transformation of an object view. According to the derivation in [83], the complex object probability, given a measurement and place, can be approximated by particles (Monte Carlo) as in Eq. 7.5.

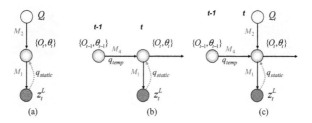

Figure 7.3. Graphical model for object recognition in video. In our combined model (c), belief at the object node (center circle) gets information from image measurement (M_1), place information (M_2), and the previous message (M_4).

$$p(X_t|z_{1:t}, Q_{1:t}) \approx \sum_{i=1}^{N} w_t^i \delta(X_t - X_t^i), \ where \ w_t^i \propto w_{t-1}^i \frac{p(z_t|X_t^i)p(X_t^i|Q_t)p(X_t^i|X_{t-1}^i)}{q(X_t^i|X_{t-1}^i, z_t)}$$

(7.5)

As you can see in Eq. 7.5, weight is updated by importance sampling where $p(z_t|X_t^i)$ represents measurement (M_1), $p(X_t^i|Q_t)$ represents scene context (M_2), and $p(X_t^i|X_{t-1}^i)$ represents temporal context (M_4) from the previous state. The importance (or proposal) function ($q(X_t^i|X_{t-1}^i, z_t)$) is defined as Eq. 7.6 This is almost the same form as introduced in [76]. The performance of the particle filter depends on modeling the proposal function (q). If we set the proposal function by using only the prior motion, the system cannot cope with the dynamic object appearances and disappearances. Therefore, we utilize the concept of mixture proposal for our problem. If q_{static} represents the proposal function defined by static object recognition and q_{temp} represents the proposal function defined by conventional prior motion, then the final proposal is defined as Eq. 7.6. Similar to place recognition, we propose three kinds of sample generation modes since we do not know the situation a priori: reinitialization (r) where $\beta = 0$, normal tracking (α) where $\beta = 1$, and hybrid tracking ($1 - r - \alpha$) where ($0 < \beta < 1$). If the temporal context is unavailable, we generate samples from only static object recognition. If in normal tracking mode, we use the conventional proposal function. If static and temporal context are available, we propose samples from hybrid (mixture) density functions.

$$p(X_t^i|X_{t-1}^i, z_t) = (1 - \beta)q_{static}(X_t^i|z_t) + \beta p_{temp}(X_t^i|X_{t-1}^i)$$ (7.6)

7.3.2 Modeling of proposal function ($q(X_t^i|X_{t-1}^i, z_t)$)

The proposal function in Eq. 7.6 consists of temporal context (prior motion) from the previous state and static context from the input image. The temporal context is modeled as Eq. 7.7. We assume that object labels and pose are independent.

$$p_{temp}(X_t^i|X_{t-1}^i) = p_{temp}(O_t^{(i)}|O_{t-1}^{(i)})p_{temp}(\theta_t^{(i)}|\theta_{t-1}^{(i)}) \tag{7.7}$$

where $O_t^{(i)} = O_{t-1}^{(i)}, t > 0$ and $\theta_t^{(i)} = \theta_{t-1}^{(i)} + u_t$. u_t is Gaussian prior motion. Temporal context means simply normal tracking state.

The static proposal function $q_{static}(X_t^i|z_t)$ is defined as a Hough transformation using corresponding local features (G-RIF) between image and model with local pose information in the features. We use the scalable 3D object representation and recognition scheme explained in the Chapter 5 using shared feature and the view clustering method in part-whole context. We can get all possible matching pairs by a NN (nearest neighbor) search in the feature library. From these, hypotheses are generated by the Hough transformation in clustered view (CFCM: common frame constellation model) ID, scale (11 bins), orientation (8 bins) space, and grouped by object ID. Then, we determine to accept or reject the hypothesized object based on the bin size with an optimal threshold [68]. Finally, we select the optimal hypotheses that can be matched best to the object features in the scene.

7.3.3 Modeling of measurement M_1

Given sample object parameters ($X_t^i = (O_t^i, \theta_t^i)$), we use a color histogram in normalized r-g space as a measurement ($M_1 = p(z_t|X_t^i)$). The model color histogram is acquired by the proposal function explained previously since the recognized object can provide an object label with object boundary information. We use χ^2 distance in the kernel recipe for measurements [113].

7.3.4 Modeling of place message M_2

The message from the place to a specific object is calculated using Eq. 7.8. We also use approximated belief propagation with the max-product rule.

$$M_2 = p(O_t^i, \theta_t^i|Q_t) \propto \max_q \{\psi(O_t^i, Q_t = q)p(Q_t = q)\} \tag{7.8}$$

Training scenes (very blurry, noisy)   Training objects (very blurry, noisy)

Figure 7.4. Composition of training and test database.

7.3.5   Modeling of temporal message M_4

The temporal message ($M_4 = p(X_t^i|X_{t-1}^i)$) in 7.6 is the same as the proposal function in 7.7. If an object particle is sampled from the temporal context only, the weight is simply the production of measurement and message from place due to the cancellation in Eq. 7.6.

7.4   Experimental Results

7.4.1   Large scale experiment for building guidance

We validate the proposed place-object recognition system in the department of Electrical Engineering building to guide visitors from the first floor to the third floor. Training data statistics and related images are summarized in Fig. 7.4 and Table 7.1 respectively. We used 120 images captured at arbitrary poses and manually labeled and segmented 80 objects in 10 places. Note that the images are very blurry. After learning, the size of the object feature was reduced from 42,433 to 30,732 and the scene features were reduced from 106,119 to 62,610. The number of learned views was reduced from 260 to 209 (2.61 views/object).

In the first experiment, we evaluate the performance of place recognition. As we said, there are four kinds of message combinations: measurement only (M_1), temporal context (HMM: $M_1 + M_4$), static context only ($M_1 + M_2$), and unified context (proposed: $M_1 + M_2 + M_4$). Fig. 7.5 shows the evaluation results for every 100 frames among 7,208. As you can see, the unified context-based place

Table 7.1. The composition of training images and test images.

Role	Scene (640 × 480)		Object	
	# of places	# of scenes	# of objects	# of views
Training	10	120	80	209
Test	10	7,208	80	12,315

recognition is better than others.

Fig. 7.6 summarizes the overall evaluation for static context only, full context (static+temporal), and temporal context only. We checked both the detection rate of objects and the relative processing time. The full context-based method shows a better detection rate than the static context only with less processing time. Fig. 7.7 shows partial examples of bidirectional place and object recognition sequences. Note that the proposed system using static context and temporal context shows successful place and object recognition in video under temporal occlusions and with a dynamic number of objects.

7.5   Summary

In this chapter, we presented modeling methods of static and temporal context focusing on bidirectional interaction between place and object recognition in video. We first modeled object to place messages for the disambiguation of places using object information. The unified context-based place recognition shows improved place recognition. We also modeled place message with measurement and temporal context to recognize objects under the important sampling-based framework. This structure of object recognition in video can disambiguate objects with reduced computational complexity. We demonstrated the synergy of the bidirectional interaction-based place-object recognition system in an indoor environment to guide visitors. The system can be applied directly to various areas for interactions between humans and computers. The proposed method can be upgraded to the category level to provide higher level place and object information to humans.

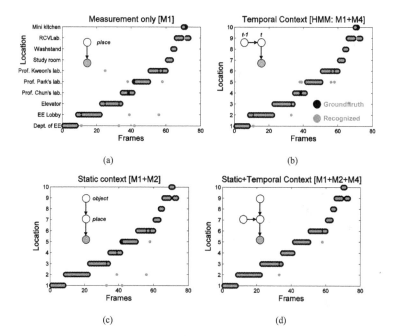

Figure 7.5. Place recognition results using (a) measurement, (b) temporal context, (c) static context, and (d) unified context for the test sequences.

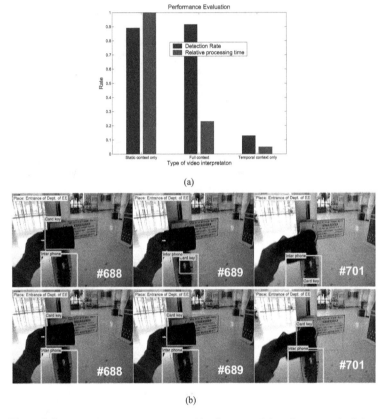

(a)

(b)

Figure 7.6. (a) Evaluation of object recognition in terms of detection rate and relative processing time, (b) (Top) Results with only static context, (bottom) using ours.

Figure 7.7. Examples of bidirectional place and object recognition results in lobby and washstand.

Chapter VIII

Pixel Context for Object Categorization

In Part I (Chapter 3-7), we focused on the robust object identification with visual contexts. In this chapter, we propose a modeling method of pixel context for novel object categorization. Visual categorization is fundamentally important for autonomous mobile robots to get intelligence such as novel object acquisition and topological place recognition. The main difficulty of visual categorization is how to reduce the large intra-class variations. In this paper, we present a new method made robust to that problem by using intermediate blurring and entropy-guided codebook selection in a bag-of-words framework. Intermediate blurring can reduce the high frequency of surface markings and provide dominant shape information. Entropy of a hypothesized codebook can provide the necessary amount of repetition among training exemplars. A generative optimal codebook for each category is learned using the MDL (minimum description length) principle guided by entropy information. Finally, a discriminative codebook is learned using the discriminative method guided by the inter-category entropy of the codebook. We validate the effect of the proposed method using a Caltech-101 DB, which has large intra-class variations.

8.1 Large intra-class variation by surface marking

Intelligent mobile robots should have visual perception capability akin to that provided by human eyes. Let's assume that a mobile robot is put in a strange house environment. It will wander the house and categorize each room as a living room, kitchen, or bathroom. Additionally, it will categorize novel objects such as the sofa, TV, dining table, or refrigerator. As we can see in this scenario, the two basic functions of an intelligent mobile robot are categorizing places and objects for automatic high-level learning about new environments. In the current state-of-the-art, topological localization remains at the level of image identification or matching for a specific environment [38, 50]. Object identification (recognition) of the same

Figure 8.1. Examples of surface markings for a cup category.

objects is almost matured due to the robustness of local invariant features such as SIFT and G-RIF [35, 54]. Currently, the categorization of general objects is an active research area in computer vision [22, 61].

However, visual categorization is a very challenging problem due to large intra-class variations. Among many sources of them, such as geometric and photometric variations, surface markings are dominant. Fig. 8.1 shows several cups. Note the various surface markings at the interior regions of the cups. The effect of surface marking is much larger in man-made objects than in animals or plants due to creative design for beauty.

To our best knowledge, there has been no work published on the reduction of surface markings in object categorization. Most researchers have focused on how to minimize intra-class variations of object shape. We can summarize the current object representation approaches as shown in Fig. 8.2. As the strength of a geometric relation is weaker, the amount of intra-class variation is smaller. At the same time, the discrimination power is reduced. PCA can represent whole objects directly and is very weak to geometric variations [48]. The constellation model of visual parts can handle geometric variations more flexibly [22, 64]. Flexible shape samples can represent large variations of shapes [7]. Bag of words, derived from document indexing, is a very robust method to visual variation because it considers no geometrical relations [13]. Texton, which is a more generalized version of bag of

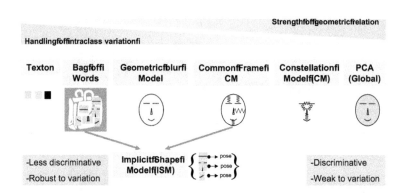

Figure 8.2. Summary of object representation schemes in terms of geometric strength and intra-class variation.

words, can categorize textured regions such as sea, sky, and forest [116]. A compromise of both extremes is the implicit shape model, which assigns pose information for each codebook [47].

In this chapter, our object representation is based on the bag of words approach to take advantage of its simplicity and robustness to large geometric variations. However, we focus on reducing surface marking problems during visual word or codebook generation. Our key idea is twofold: One is to apply intermediate blurring to extract important object shape information. It is motivated from cognitive experiments showing that human visual systems can categorize blurry objects very quickly [4]. The other is based on information theory. Entropy of a hypothesized codebook among training instances should be high for surface marking reduction, and entropy among different categories should be low for discrimination.

8.2 Background of bag of visual words

The term visual words originated from linguistics [16]. A paragraph consists of a set of words. Likewise, we can think of a scene or an object as composed of visual words, as shown in Fig. 8.3. Recently, the bag of visual words approach has shown very promising results on visual categorization problems [13,14,61,116]. Although it is a very simple representation, it can handle large geometric variations because

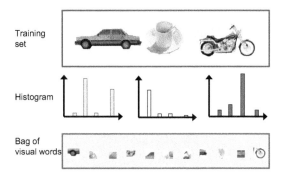

Figure 8.3. The bag of visual words-based object representation scheme.

it discards geometric relationships among features. The basic steps for the bag of visual words approach are visual word generation, histogram building, and classifier learning. The key issue of the visual word-based classification is how to learn the optimal set of visual words, or codebook. Csurka et al. selected the optimal set of visual words by k-means clustering [13]. The size of k is empirically selected by cross validation of the training set. Winn et al. proposed a pair-wise feature clustering method that maximizes inter-class variation and minimizes intra-class variation [116]. Previous approaches do not consider surface marking problems for optimal codebook generation.

8.3   Robust categorization to surface markings

8.3.1   Overall categorization system

The proposed object categorization system is composed of feature extraction, codebook generation, and classification, as shown in Fig. 8.4. First, we extract dense features after intermediate blur. Then an intra-class codebook is learned using the model selection method of entropy-guided MDL (minimum description length) as the intra-class training set. These intra-class codebooks are further learned in a discriminative way using entropy-guided codebook selection as the inter-category training set. Then each training instance is represented by histogram using the opti-

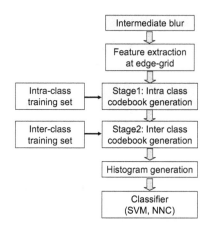

Figure 8.4. Overall categorization system for surface marking robustness. Intermediate blur and stage 1 blocks provide key roles for the reduction of surface markings.

mal codebook learning. Finally, classification is conducted using either SVM (support vector machine) or NNC (nearest neighbor classifier) by varying distance metrics. The most important blocks for surface marking reduction are intermediate blur and stage 1. Details of the system are explained in the following sub-sections.

8.3.2   Information theoretic parameter selection for features

The first issue in the bag of visual words approach is how to extract local features. Direct application of sparse scale invariant features such as SIFT [54] and G-RIF [35] to Caltech-101 DB (available at http://www.vision.caltech.edu/htmlfiles/archive.html) shows very disappointing results: a 26.8% correct classification rate using 15 images for training and 15 images for testing (using Berg's evaluation method [7]). So, we need to find an optimal set of feature parameters, such as level of blur, location of sampling points, size of region, and image scale.

Motivated from the basic constraint (maxVar(inter-class)/Var(intra-class)) and entropy in information theory [88], the codebook (F) should have high entropy ($H(C|F)$) within a category , and low entropy among categories. For the evaluation, we used a PCA-GRIF descriptor (5-dim) and calculated entropy in a partitioned feature space. For a given partition $A = \{A_i\}$, entropy is $H(A) = -\sum_i p_i \log p_i$ where p_i is the relative bin count. We set the partition as 10/cell and used 10 selected categories. Due to the properties of the Caltech DB, we set the scale as fixed. The final parameter is selected at the value where the difference of inter-category entropy and intra-category entropy is maximized. Fig. 8.5 shows the evaluation results. According to the maximum value, we set the blurring level as $\sigma = 4\sqrt{2}$, region radius as 26 pixels, and the sampling interval at 20 pixels.

Finally, we also evaluated the edge sample and dense grid sampling types with the optimal feature parameters. The evaluation was conducted using conventional k-means clustering for codebook generation and bag of visual words framework with an NNC classifier for segmented objects. The edge-grid sampling shows upgraded performance as shown in Fig. 8.6. So, we used edge-grid sampling with the selected feature parameters.

8.3.3   Stage1: Intra class codebook generation (generative)

In stage 1, we have to minimize intra-class variations. The main cause of large intra-class variation is surface markings, which have various patterns for object instances. As shown in Fig. 8.7, the surface markings can be removed by finding repeatable parts or high-entropy parts.

Based on this relation of entropy and surface markings, we can conduct model selection using MDL more efficiently. The MDL criteria can provide an optimal codebook in terms of fitting distortion and model complexity, as shown by Eq. 8.1 [105]. The key point for surface marking reduction is to remove codebook candidates that have low entropy as shown in Fig. 8.8. An initial codebook is generated using two steps of agglomerative clustering (bottom-up) and k-means clustering (top-down) [36]. The detailed algorithm for intra-class codebook selection is shown in Algorithm 8.1.

$$\hat{\Lambda}(\mathbf{X}, \theta) = \arg\min \left\{ -\log L(\mathbf{X}, \theta) + \frac{K(V+1)}{2} \log N \right\} \qquad (8.1)$$

where L is likelihood of data fitting, $\mathbf{X}$ is training features, θ is parameters (mean and var for codebook), K is the number of codebook, V is the number of parameters per codebook, and N is the number of features. Fig. 8.9 shows the MDL model selection curve and the properties of the selected codebook. Note that our codebook can find semantically meaningful parts for the training instances regardless of various surface markings.

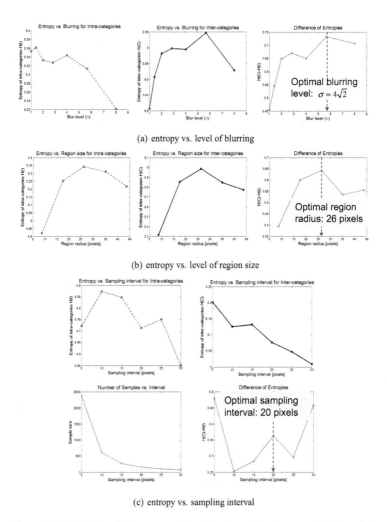

(a) entropy vs. level of blurring

(b) entropy vs. level of region size

(c) entropy vs. sampling interval

Figure 8.5. Evaluation of feature parameters (blurring, region size, sampling interval) in terms of intra- and inter-category differences of entropy.

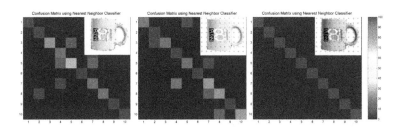

Figure 8.6. Evaluation (confusion matrix) of sampling type: (left) edge sampling, (middle) grid sampling, (right) edge-grid sampling. Edge-grid sampling shows better performance.

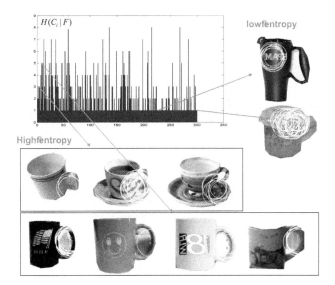

Figure 8.7. Observation for repeatable parts (high entropy) and surface marking parts (low entropy).

Algorithm 8.1 Class-specific codebook generation

```
Given:  category-specific local features
Goal:  make codebook
```
Step 1. Extract edge-grid features for each category.
Step 2. Make initial codebook using appearance-based clustering [36].
Step 3. Starting from this initial codebook.
 Evaluate MDL (Eq. 8.1)
 If MDL is minimum, stop.
 Else
  Remove one codebook that has lowest entropy. Go to 1.

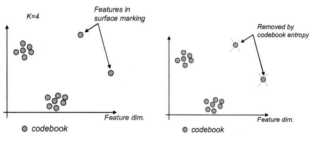

(a) concept of k-mean clustering   (b) concept of entropy-guided codebook

Figure 8.8. The mechanism of surface marking removal in entropy guided codebook compared to the conventional k-means clustering.

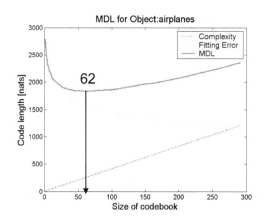

(a) MDL graph for airplane category

(b) examples of selected optimal codebook

Figure 8.9. Entropy-guided MDL graph and its example parts corresponding to selected codebook. Note that similar parts are selected regardless of surface markings.

8.3.4 Stage2: Inter class codebook generation (discriminative)

Given the category-specific codebooks learned in stage 1, we have to select a discriminative universal codebook for bag of visual words-based classification. We can obtain a discriminative codebook (F_{opt}) by maximizing the posterior of class labels given training examples and a hypothesized universal codebook. The key point in this approach is to select a removable codebook using the inter-category entropy of a codebook that has large entropy (ambiguous codebook). If we define $\{F\}$ as a hypothesized universal codebook, I_i^c as the i-th object instance belonging to category c, and l as the category label, then the posterior can be formulated as Eq. 8.2.

$$F_{opt} = \arg\max_F \left\{ \prod_c \prod_{i \in c} p(l|I_i^c, \{F\}) \right\}$$

$$= \arg\max_F \left\{ log \left(\prod_c \prod_{i \in c} p(l|I_i^c, \{F\}) \right) \right\}$$

since

$$p(l|I_i^c) = \frac{p(I_i^c|c,\{F\})p(c,\{F\})}{\sum_{c'} p(I_i^c|c',\{F\})p(c',\{F\})}, \; assume \; uniform \; p(c,\{F\}) \qquad (8.2)$$

$$F_{opt} = \arg\max_F \left\{ \sum_c \sum_{i \in c} \left(\log p(I_i^c|c,\{F\}) - \log \sum_{c'} p(I_i^c|c',\{F\}) \right) \right\}$$

$$where \; p(I_i^c|c,\{F\}) = p(H_i^c|H_M^c) \propto \exp\left(-KL(H_i^c, H_M^c)\right)$$

$$and \; KL(H_i^c, H_M^c) = \sum_{j=1}^{|F|} \left(H_i^c(j) - H_M^c(j) \right) \log \frac{H_i^c(j)}{H_M^c(j)}$$

The posterior criterion in the 4th line of Eq. 8.2 is used to select the optimal set for a discriminative codebook. Fig. 8.10 shows the codebook search algorithm and its learning graph. Fig. 8.11 shows the test results using only a set of the intra-class codebook ($|F| = 1062$) and the discriminatively learned universal codebook ($|F| = 926$) for 10 object categories.

8.3.5 Distance metrics and classification

After histogram building from the discriminative codebook for all the training instances, we have to learn classifiers with certain distance metrics. We can summarize these as follows.

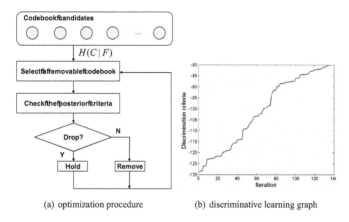

(a) optimization procedure   (b) discriminative learning graph

Figure 8.10. Inter-category entropy-guided universal codebook selection method.

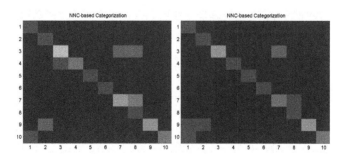

Figure 8.11. Confusion matrix using non-discriminative codebook and discriminative codebook. (left) Before discriminative learning, (right) after discriminative learning.

Distance metrics: $D(H_t, H_m)$
- Euclidean distnace: $D(H_t, H_m) = \sum_i (H_t(i) - H_m(i))^2$
- KL-divergence: $D(H_t, H_m) = \sum_i (H_t(i) - H_m(i)) \cdot \log (H_t(i)/H_m(i))$
- Intersection: $D(H_t, H_m) = \sum_i \min (H_t(i), H_m(i))$
- χ^2 distance: $D(H_t, H_m) = \sum_i \frac{(H_t(i) - H_m(i))^2}{H_t(i) + H_m(i)}$

Classification

- NNC is the simplest classifier because it requires no specific learning method. Each training histogram is regarded as a single prototype. So, for an unknown test histogram, a category label is assigned with the nearest prototype in the model database.

- Support vector machine (SVM) [107] can learn classification boundaries from training samples. It has been the most powerful classifier until now. Recently, a kernel-based SVM was introduced that can learn non-linear classification boundaries for complex data. In the extended Gaussian kernel, we can use the distance metrics described above. In the experiment section, we will compare these classification methods using codebooks that are robust to surface markings and discriminative.

8.4 Performance evaluation

8.4.1 Fixed scale sampling

We evaluated our categorization system using a Caltech-101 DB. It consists of 48 man-made objects and 53 animals and plants. In initial experiments, we use the sampling parameters selected in previous section and we extend to multi-scale samplings. We selected 1 human face and 9 man-made objects, such as airplanes, cameras, cars, cell phones, cups, helicopters, motorbikes, scissors, and umbrellas, which have large intra-class variations due to surface markings. We randomly selected 15 examples for each category and tested 15 unlearned cluttered examples. The first experiment was conducted using codebooks obtained from stage 1 learning (GC: generative codebook) and stage 2 learning (DC: discriminative codebook). Table 8.1 summarizes those experiments. In this test, NNC with KL-divergence showed the best classification results.

Based on this finding, we extended the experiment to the whole database. We selected the NNC classifier with KL-div. distance. The DC was learned from each category-specific GC. The average classification of our system was 48.58% for a

Table 8.1. Summary of classification evaluation in terms of distance metrics, type of codebook using NNC.

CB type	No. of CB	Euclid	KL-div	Intersec.	χ^2
GC	950	75.3%	79.3%	78.0%	77.3%
GC	513	70.6%	77.3%	76.6%	78.0%
DC	348	66.0%	**81.3%**	77.3%	74.0%

cluttered test set as shown in Fig. 8.12.The current state-of-the-art for the same database using the bag of visual words (single level, L=0, 15 training) shows 41.2% [44]. Most incorrect classifications are for animals and plants.

8.4.2 Extension to multi-scale sampling

G-RIF vs. SIFT: As a starting point to multi-scale feature selection, first we compare our G-RIF to SIFT for the 10 category classification problem. For fair comparison, we make codebook using the conventional k-means clustering with k=500. We use NNC classifier with χ^2 distance. 15 images per category are used for training, rest 15 images are used to test. Fig. 8.13 shows the confusion matrices. The average categorization rate of SIFT is 74% and that of G-RIF is 81.3%. Since G-RIF is generalized version of SIFT in terms of information quantity, bag of visual words using G-RIF shows better performance than the same method using SIFT.

Entropy-guided codebook vs. k-means clustering: Next, we check the effect of codebook selection (our entropy-guided codebook, k-means clustering) to the categorization performance. We use the same G-RIF feature, number of codebook, and classifier except different codebook generation method. As shown in Fig. 8.14, bag of visual words using entropy-guided codebook shows better performance the the same method using k-means clustering. This fact means that our codebook can handle surface markings more robustly.

Optimal distance measure and classifier: Until now, we select G-RIF and entropy-guided codebook. Then what is optimal distance and classifier for the codebook histogram? We compared 4 distance measures of Euclidean, KL-divergence, histogram intersection, and χ^2 distance. As classifiers, we compare the well-known NNC and SVM especially OvR (one versus rest) for multi-category classification.

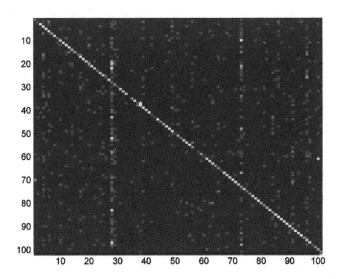

Figure 8.12. Extended experiment for the whole Caltech-101 DB using NNC-KL div. classifier with DC.

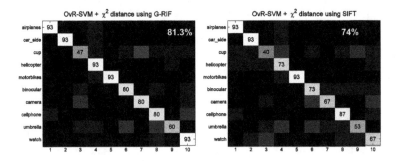

Figure 8.13. Comparison of G-RIF and SIFT for 10 Caltech DB.

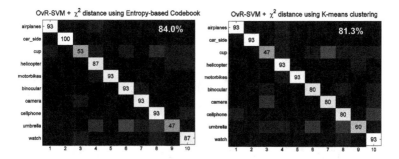

Figure 8.14. Comparison of our entropy-guided codebook and k-means clustering.

Table 8.2. Categorization rate [%] according to the combinations of distance measures and classifiers

distance	Euclidean dist.	KL-divergence	histogram intersection	χ^2
NNC	51.3	68.6	68.0	70.0
SVM	80.0	81.3	86.0	86.0

Fig. 8.15 shows the confusion matrices for the combinations of distances and classifiers. Table 8.2 summarizes the overall performance. Note that the histogram intersection or χ^2 distance with SVM is the optimal choice for bag of visual words method.

Sparse sampling vs dense sampling in multi-scale: We have determined the optimal codebook selection method and classifier for bag of visual words-based categorization. In previous section, we uses dense sampling with fixed scale. Now, we compare the sparse sampling and dense sampling in multi-scale space. We set all the related parameters to the same value except sampling method in scale space. As shown in Fig. 8.16, the dense sampling in scale-space shows much better performance (87.3%).

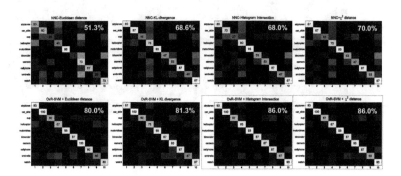

Figure 8.15. Comparison of distance measures and classifiers.

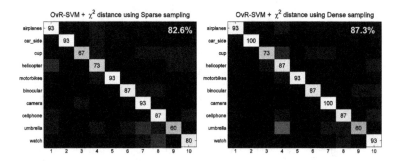

Figure 8.16. Comparison of sparse sampling and dense sampling.

8.5 Summary

In this chapter, we presented an object categorization method focusing on surface markings in the bag of visual words framework. We can minimize the effect of surface markings based on the entropy of the codebooks. High entropy in the intra-class codebook can remove surface marking parts (low entropy) in stage 1 learning. Additionally, a discriminative codebook is also selected from the category-specific codebook guided by the entropy of the inter-class codebook. The high entropy codebook is removed first because it gives ambiguous class labels. Finally, we evaluated those codebooks using NNC and SVM classifiers with different distance metrics. With the optimal set of features, codebooks, and classifiers, we can get upgraded performance in the bag of visual words framework. This work for codebook selection and classification is applied to more complex categorization method in Chapter 9.

CHAPTER IX

PART-PART, PART-OBJECT CONTEXT FOR SIMULTANEOUS CATEGORIZATION AND SEGMENTATION

Chapter 8 introduced an entropy-guided codebook selection method for object categorization. Based on this work, we present a simultaneous object categorization and segmentation method. According to our experience, the performance of object categorization for cluttered image is degraded enormously due to ambiguous figure-ground. We solve such problem by utilizing three kinds of visual context: part-part context in bottom-up hypothesis, part-object context, and object-background context additionally. We integrate the object categorization and segmentation with a generative framework where three kinds of contexts are incorporated. We choose optimal category labels with figure-ground masks that can best describe input features. Several experiments validate the robustness for cluttered general images.

9.1 Background clutter in categorization

Currently, object categorization is highly active topic in computer vision. The definition of object categorization is to assign a category label (normally basic level) for a novel object. The main problem of object categorization is large intra-class variations due to photometric and geometric changes. Pioneering work for this research is constellation model proposed by Fergus et al. [22]. It can handle visual variations with part-based spring model. Recently, bag of keypoints method shows more robustness to geometric variations since it discards geometric relations among parts [13,116]. Based on the bag of keypoints, extended methods are proposed such as spatial pyramid [44], hyperfeatures [1], and sparse localized features [69] that encode spatial information to histograms. Zhang et al. focus on classifier rather than feature extraction [121]. They combine nearest classifier with SVM, called SVM-KNN. It shows much upgraded categorization performance to the Caltech-101 DB

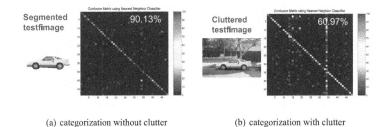

(a) categorization without clutter   (b) categorization with clutter

Figure 9.1. The effect (confusion matrices) of background clutter to object categorization using the bag of keypoints.

(66.23%).

These methods assume objects as whole images so it is very similar to image classification. If there are background clutters, they regard the clutters as parts of objects during learning. If we learn objects without background clutter and test two sets of images (segmented, cluttered) using bag of keypoints, we can get important results as shown in Fig. 9.1. This confusion matrices represent the object categorization for 48 man-made objects of Caltech DB. Note that the categorization accuracy is degraded from 90.13% to 60.97% (almost 30%).

Recently, several researchers have tried to reduce background clutter problems in object categorization. In feature level, feature selection [14], or boosting [77] is proposed to overcome the clutter problem. Leibe et al. proposed combined object categorization and segmentation with implicit shape model (ISM) [47]. First they estimate object category and then segment figure-ground in pixelwise. Spatial relation is modeled in a maximum entropy framework and lead to high categorization rate [43]. Direct object region detection using boundary fragment, similar model to ISM, is also proposed and shows some promising results to cluttered objects [78,93]. Partial matching method such as χ^2 distance can alleviate background clutter during categorization using SVM [28]. Object segmentation with given category information using random field model shows good segmentation results even for occluded objects [117]. However, this does not solve the categorization problem.

All the approaches try to solve background clutter problem in terms of object categorization or object detection (localizing objects given a category). These methods are partial solutions to our goal, categorizing and segmenting of unknown objects. We attack the background clutter problem by actively utilizing spatial context (Fig. 9.2(a), 9.2(b)) and hierarchical context (Fig. 9.2(c)). Part-part context

(a) part-part context   (b) object-background context

(c) part-object context

Figure 9.2. Visual context-based strategy for background clutter problem

means that parts belonging to the same object category should have same property. Gestalt's law of proximity and similarity for part-part context can provide group of parts as shown in Fig. 9.2(a). Parts belong to background region rarely show clustering property compared to parts in object region. Object-background context means that objects usually have familiarity to certain background. Fig. 9.2(b) shows car examples with street background. If the relation between object and background is stronger, then we can categorize unknown object more accurately. The most important context is part-object relationship as shown in Fig. 9.2(c). A part can predict object category, viewpoint and figure/ground mask. Likewise, whole object can verify category existence and object region by carefully checking related parts.

These context are modeled by a directed graphical model which can provide object category with figure-ground segmentation. Bottom-up evidence from part-part context and part-object context can provide proposal function. Top-down generative inference using object-background context and object-part context can provide optimal category label, viewpoint, and figure-ground mask which can best describe input features (both object and background features). The inference is conducted by multi-modal MCMC sampling. Experimental results validate the power of the proposed framework for object categorization and figure-ground segmentation in

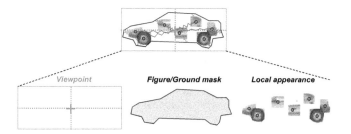

Figure 9.3. Joint appearance and shape representation of an object. It consists of viewpoint, figure-ground mask, and local appearance.

cluttered environment.

9.2   Examplar-based category representation

9.2.1   Joint appearance and shape model

To fully utilize the visual contexts, we propose a joint appearance and shape model with a viewpoint of an object instance as shown in Fig. 9.3. In general, viewpoint can be 3D view point (rotation, translation) for 3D object. But we restrict it to object center with scale in this work. A figure-ground mask divides an image into figural region and background region. Finally, local patches represent part-based object appearance. The viewpoint, figure-ground, and local appearance are interrelated like spring model. In this joint model, local patches have important role since they relate viewpoint and figure-ground boundary. If we know a part, then we can predict viewpoint and object boundary. This is the part-object context explained in the previous section.

9.2.2   Extension to category representation

We represent a category by extending the joint appearance and shape model as shown in Fig. 9.4. Local appearances are represented in terms of category specific codebook. Each examplar contains viewpoint, figure-ground mask, and poses (relative position to viewpoint center, characteristic patch scale) of local patches.

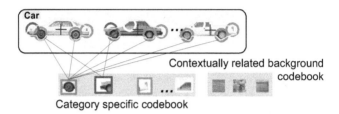

Figure 9.4. Examplar-based category representation using joint appearance and shape model

There is also background codebook which is contextually related to a certain category. Details of modeling and learning will be explained next sections.

9.3 Directed graphical model for object categorization and segmentation

9.3.1 Basic concept and directed graphical model

Look at the object in cluttered environment as shown in Fig. 9.2. We can generate such images if we have category label, viewpoint of an object, figure-ground region, and codebook corresponding to input features. Fig. 9.5(a) shows such example of generative procedure. We assume a single object in cluttered background since it is basic block for multiple object categorization. The parameter $\{C, B\}$ represents a pair of category label C and related background label B. Given $\{C, B\}$, first we can generate viewpoint of an object. If the object is 3D, then the viewpoint is rotation and translation in 3D world. Currently, we focus on image-based object representation. So the viewpoint parameter V contains viewpoint center (x_c, y_c) and object scale factor (s) relative to model size. In the next layer, figure-ground mask (M) is generated from category-background information and viewpoint. Mask M consists an array of $\{0, 1\}$ where 0 represents background pixel and 1 represents foreground pixel. In the third layer, codebook index F is selected using category-background information and figure-ground mask. Finally, we can generate input features G using the selected codebook and viewpoint information. G consists of

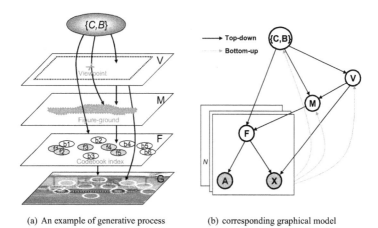

(a) An example of generative process   (b) corresponding graphical model

Figure 9.5.   Generative framework for simultaneous object categorization and figure-ground segmentation in cluttered environment.

N local appearance A and part pose X. Viewpoint information is reflected to part pose generation.

Fig. 9.5(b) shows the directed graphical model (Bayesian Net) exactly corresponding to Fig. 9.5(a). White nodes represent hidden variables and shaded nodes represents observed variables. Note the causal relationship between nodes. Due to the N input features, we replicate the codebook index and observation nodes N times (boxed regions). In addition to the top-down generative model, we draw bottom-up (dotted arrow) flow for fast estimation.

9.3.2   Mathematical formulation

Given an unknown object with cluttered background, we can detect multi-scale input features $G = \{g_i = (a_i, x_i)\}, i = 1, 2, \cdots, N$. Assume that we already have trained model D which has labeled, segmented, and viewpoint with learned parameters (Learning will be explained in the next section). Then, the object categorization and segmentation problem is to estimate category label C, figure-ground mask $M(i, j) = 1$ or 0, and viewpoint $V = \{x_c, y_c, s\}$. We set the solution vector as $H = (C, M, V)$ and the solution space as Ω. Then the optimal solution is expressed as Eq. 9.1. Since we maximize the posterior, the normalization is omitted.

$$
\begin{aligned}
H^* &= \arg\max_{H \in \Omega} p(H|G, D) \\
&= \arg\max_{H \in \Omega} p(G|H, D)p(H|D)
\end{aligned}
\tag{9.1}
$$

According to the graphical model (see Fig. 9.5(b)), the prior term $p(H|D)$ is decomposed into three conditional probability as Eq. 9.2. From training data D, $p(C|D)$ represents the prior of category label. Given category label C and D, $p(V|C, D)$ represents the prior of viewpoint. Given a category, viewpoint with data, we can generate figure-ground mask M from $p(M|C, V, D)$.

$$
p(H|D) = p(C|D)p(V|C, D)p(M|C, V, D)
\tag{9.2}
$$

Given a hypothesis $H = (C, V, M)$ and trained data D, the likelihood term $p(G|H, D)$ is defined as Eq. 9.3.

$$
p(G|H, D) = p_f(G_f|H, D)p_b(G_b|H, D)
\tag{9.3}
$$

where $G_f = \{g_m : M(x_m) = 1\}$, $G_b = \{g_n : M(x_n) = 0\}$. x_m is the position of input feature g_m in image space. Each likelihood term is defined as Eq. 9.4.

$$
\begin{aligned}
p_f(G_f|H, D) &= \prod_{i=1}^{N_f} \left(\sum_{j=1}^{|F_f|} \phi_j N(a_i; \mu_a^j, \Lambda_a^j) N(x_i; s \cdot \mu_x^j + (x_c, y_c), \Lambda_x^j) \right) \\
p_b(G_b|H, D) &= \prod_{i=1}^{N_b} \left(\sum_{j=1}^{|F_b|} \phi_j N(a_i; \mu_a^j, \Lambda_a^j)/A \right)
\end{aligned}
\tag{9.4}
$$

where N_f is the number of input features generated by object codebook F_f and N_b is the number of input features generated by background codebook F_b. So $N_f + N_b = N$ which is the total number of input features. ϕ_j is the probability of codebook j. Foreground features are generated by Gaussian distributions. Note that the mean of codebook is affected by the viewpoint $V = (x_c, y_c, s)$. Background features are generated by background codebook. However, the pose distribution is uniform since they are distributed randomly in area A. Details of learning and inference will be explained next sections.

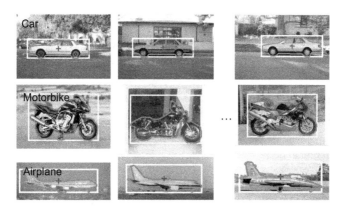

Figure 9.6. Labeled object examples for training. For each object category instance, figure-ground is assigned by manual segmentation.

9.4   Examplar-based Learning

9.4.1   Prior for category, viewpoint, and mask

Prior distributions in Eq. 9.2 are learned using a set of labeled training images as shown in Fig. 9.6. Let database D have category label C_{DB}, viewpoint V_{DB}, and figure-ground mask M_{DB} for each instance indicated in Fig. 9.6. At this state, parameters related to codebook (ϕ, μ, Λ) are null.If there are N_C categories and each category has N_M examples, then the category prior $p(C|D)$ is uniform as $1/N_C$. Given a category, the viewpoint distribution can be estimated directly from labeled examples. However, we define $p(V|C, D)$ as Eq. 9.5 for the generalization. In real environment, objects can be anywhere in an image. We restrict the scale factor in the range of $[0.5\ 2]$. Given category label, viewpoint, and figure-ground masks in D, the prior $p(M|C, V, D)$ is defined as $1/N_M$ since we randomly choose figure-ground mask in the database.

$$p(x_c, y_c, s|C, D) = \frac{1}{A} \cdot \frac{1}{1.5} \qquad (9.5)$$

9.4.2 Entropy-based codebook learning

For the likelihood estimation in Eq. 9.4, we have to learn parameters related to code-book. The required parameters are probability of codebook appearance ϕ, means and variances of codebook descriptor (both foreground and background codebook), and that of codebook pose (foreground). In codebook selection, we use the statistical properties explained in Chapter 8 rather than simple k-means clustering. The entropy-guided codebook represents repeatable or semantically meaningful parts removing surface markings.

Basically almost the same procedures are used to choose codebook but we modify it to fit the joint appearance and shape model. The learning is conducted category-wise since our systems is generative framework. First, we extract dense (or sparse) features in scale-space from foreground regions as show in Fig. 9.7. Positions of local features are defined in polar coordinates centered on viewpoint. Assume that we have finite (ex. 15) images. Through agglomerative clustering and k-means clutering, we can obtain candidate codebook F_{hyp}. For each codebook candidate F, we can estimate entropy of instance label L as Eq. 9.6. $p(l|F)$ is the relative frequency of codebook F in object instance l. Fig. 9.8 shows the entropy of candidate codebook. Codebook whose entropy is low belong to non-repeatable parts such as surface markings (see the parts in FEDEX) or distinctive parts. A codebook belonging to the wheel parts shows high entropy and represents semantically meaningful.

$$H(L|F) = -\sum_{l \in L} p(l|F) log_2 p(l|F) \tag{9.6}$$

In Chapter 8, we selected optimal codebook using MDL (minimum description length) guided by the entropy information. In this section, candidate codebook is first filtered by entropy values because we also have to consider the statistical property of pose for each codebook. We select codebook candidates whose entropies are larger than the entropy threshold (0.5, empirically tuned). Based on such candidate codebook, we check the entropy of codebook pose. In our joint appearance and shape model, appearance codebook is very important to predict viewpoint and figure-ground mask. More stable of part position, more accurate estimation is obtained. If we quantize part position in image space and part scale in scale-space as shown in Fig. 9.9, we can estimate the probability of part position $p(\chi|F)$ for codebook F. Likewise, we can estimate the probability of par scale $p(\sigma|F)$. Positional entropy and scale entropy is calculated using Eq. 9.6. Fig. 9.9 (left) shows

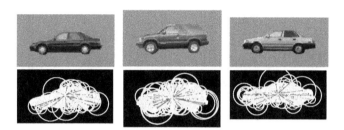

Figure 9.7. Foreground objects and detected local features. Due to dense features, every fifth features are displayed.

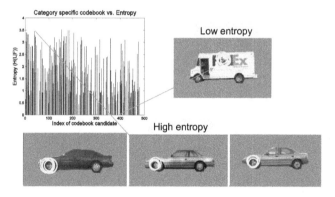

Figure 9.8. Entropy of candidate codebook appearance and examples of low entropy codebook and high entropy codebook.

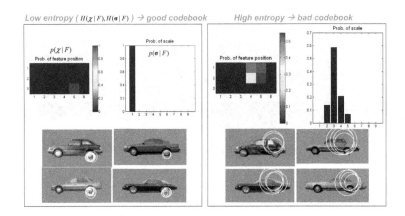

Figure 9.9. Probability distribution of codebook pose and entropy. We select codebook whose positional entropy is low.

codebook whose pose entropy (uncertainty) is low and Fig. 9.9 (right) shows codebook whose pose entropy is high. Final codebook is selected by thresholding pose entropy. We choose codebook whose postion entropy and scale entropy are smaller than 1.5 (empirically tuned). The pose entropy very meaningful to model object categories. If the pose entropy is high for all codebook, then our joint appearance-shape model is not suitable since objects usually have textured (repeated pattern) surfaces. In such case, conventional bag of keypoints [13]-based category representation is more suitable.

Through the entropy-based codebook selection, we can get final codebook. The codebook parameters for appearance is estimated by sample mean (μ_a) and sample variance (Λ_a). For simplicity, we consider only diagonal variance. The parameter estimation of codebook pose is rather difficult since a codebook can be positioned on different locations. A Gaussian mixture model can represent such phenomenon but the complexity of learning increases. We model codebook pose by compromising non-parametric and parametric representation scheme as shown in Fig. 9.10. For each examplar, sample mean and sample variance of codebook pose is estimated in polar coordinates from clustered features (see the enlarged image). The sample mean is $\mu_x = (\hat{\chi}, \hat{\sigma}) = ((r, \theta), \hat{\sigma})$. This process is repeated to other examplars where the codebook belongs to. We assume uniform distribution of examplars.

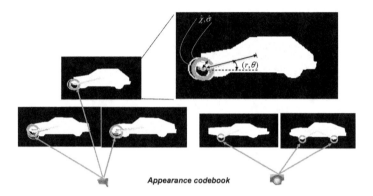

Figure 9.10. Modeling of codebook pose information using examplars.

Fig. 9.11 represents partial examples of codebook for each instance. Every third codebook is overlaid to discern different codebook.

The parameter estimation (μ_a, Λ_a) for background codebook is almost the same as the foreground codebook except the codebook pose. As we said, we assume that the pose of background codebook is randomly distributed in image space and scale-space.

9.5   Inference by boosted Markov Chain Monte Carlo

We can get optimal object categorization and figure-ground segmentation by solving Eq. 9.1. Due to the high dimensionality, direct inference is intractable. We utilize approximate inference method using sampling method such as Markov Chain Monte Carlo (MCMC) [27]. MCMC samples guarantees to converge to the posterior distribution. Metropolis-Hastings (M-H) algorithm is often used for the MCMC inference. The original MCMC can provide globally optimal solution with the cost of long time (many samples). We utilize the M-H sampling but we modify the proposal function $(q(H \rightarrow H'))$ by multi-modal distribution. It consists of prior distribution and boosted distribution from bottom-up inference (see the dotted arrows in Fig. 9.5(b)). Samples from multi-modal distribution are accepted with probability α defined as Eq. 9.7. Fig. 9.12 shows the overall inference flow graphically. Details

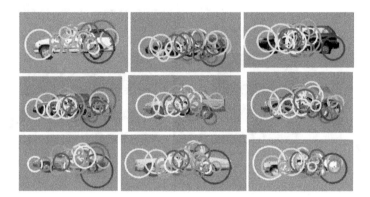

Figure 9.11. Examples of learned codebook overlaid on examplars. Different color represents different codebook.

of bottom-up proposal and multi-modal sampling-based inference are explained in the following subsections.

$$\alpha = \min \left\{ 1, \frac{p(H'|G,D)}{p(H|G,D)} \cdot \frac{q(H' \rightarrow H|G,D)}{q(H \rightarrow H'|G,D)} \right\} \qquad (9.7)$$

9.5.1   Bottom-up proposal driven by context-based boosting

Two-layered codebook: In the previous learning section, we introduced category-specific codebook generation and examplar based pose encoding scheme as shown in Fig. 9.4 and Fig. 9.10. For multiple object categorization and segmentation in a bottom-up way, we extend the category specific codebook as Fig. 9.13. It has two-layered codebook structure. From a set of category-specific codebook (CC), universal codebook (UC) is built by clustering based on appearance similarity. In Fig. 9.13, wheels in car codebook and in airplane codebook have similar appearance. Each UC contains all possible link information to CC. This link information is useful to bottom-up inference for an unknown object.

Dense feature clustering using similarity and proximity: We extract local features at dense points such as edge samples and random samples as shown in Fig.

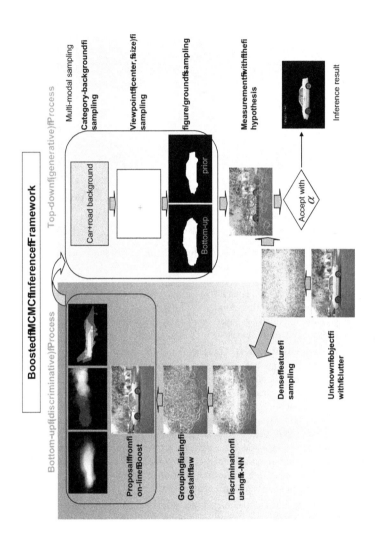

Figure 9.12. Boosted MCMC-based inference framework for simultaneous object categorization and figure-ground sementa-tion.

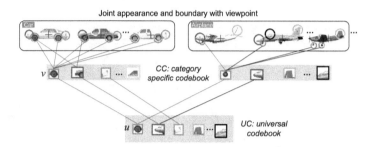

Figure 9.13. Composition of two-layered codebook (universal codebook + category specific codebook) for bottom-up inference.

9.12 (bottom). The average number of features per 320×240 image is 1000. It is not efficient to use directly such huge number of features for bottom-up inference. Instead, we filter out the dense features using discrimination by k-NN (nearest neighbor) classifier with UC. Then filtered dense features are grouped according to similarity of appearance and proximity. Similar features within 25 pixels are grouped. We denote the finally grouped features as **e**. Left middle image in Fig. 9.12 show the clustered features whose color represents the index of UC.

On-line boost using visual context: Given evidence (**e**, clustered from dense features), we can directly estimate the proposal function in bottom-up way using two kinds of visual context. The first context is *part-object* relation which is a sort of hierarchical context. An evidence e_k can predict a codebook in UC. Since UC contains CC links, we can predict category (C), viewpoint (V), and figure-ground mask (M). Fig. 9.13 and Fig. 9.10 will help you to understand the part-object prediction mechanism. The second context is part-part relation. As shown in Fig. 9.14, the quality of current interesting evident e_k is affected by neighboring evidences $N(k)$. We can predict viewpoint of e_k using part-object context. Neighboring evidences can also provide viewpoints. If these viewpoints are compatible to the viewpoint by e_k, then we accept the prediction of current evidence.

Based on the concept of visual contexts, we can model this phenomenon mathematically by borrowing the concept of boosting [110]. In the original boosting, a strong classifier (g) is constructed from a set of weak classifier (h_k) as Eq. 9.8. The weak classifier weight α_k is learned off line using positive and negative training examples.

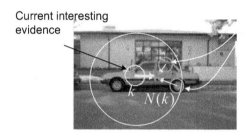

Figure 9.14. Concept of part-part context. The quality of current interesting evidence is determined by neighboring evidences.

$$g(x) = \sum_{k=1}^{k_{max}} \alpha_k h_k(x) \qquad (9.8)$$

The joint category and viewpoint classifier $g(C, V, M|\mathbf{e})$ is defined Eq. 9.9. Given an input evidence e_k, we can predict category (C), viewpoint (V), figure-ground mask (M) using part-object context such as evident to UC, UC to CC, and CC to examplar in DB. L_i is all possible interpretation links. We assume $p(C, V, M|I_i)$, $p(I_i|e_k)$ uniform for simplicity. The part-part context is utilized to estimate the weight α_k of weak classifier (parenthesis in Eq. 9.9). Compared to the conventional off-line learning α, it is learned on-line using neighboring evidences. So we call our bottom-up inference as on-line boost. The α_k for the weak classifier is defined as Eq. 9.10.

$$g(C, V, M|\mathbf{e}) = \sum_{k=1}^{k_{max}} \alpha_k \left(\sum_i p(C, V, M|L_i)p(L_i|e_k) \right) \qquad (9.9)$$

$$\alpha_k = \frac{n_{support}}{|N(k)|} \qquad (9.10)$$

where $n_{support}$ is the number of support from evidences $N(k)$. We increase the support if $|V(k) - V(j)| < \delta$ where $j \in N(k)$. Empirically, we can obtain good

With neighbor support   Without neighbor support

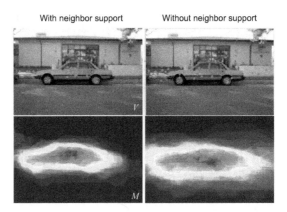

Figure 9.15. The effect of part context in bottom-up estimation. Left column shows the viewpoint and figure-ground mask boosting using part context and right show that without part context (all $\alpha_k = 1$).

estimation if we quantize the α_k. We set $\hat{\alpha}_k = 1$ if $\alpha > 0.5$, otherwise, $\hat{\alpha}_k = 0$. This can remove outlier robustly. Fig. 9.15 shows the effect of part context in bottom-up boosting. Note that the role of part context in on-line boosting of category, viewpoint, and figure-ground mask. This on-line boosting is very similar to voting in (C, V, M) space. With this bottom-up inference method, we compare dense sampling (random + edge samples) and sparse sampling (Harris+DoG points) in scale space.

Fig. 9.16 shows an example of bottom-up boosting with two kinds of sampling. Dense sampling-based boosting shows more stable evidence. Fig. 9.17 shows the robustness to scale changes in bottom-up boosting. In this small set of test, we can conclude that our part-context, dense sampling in scale-space are important to get stable bottom-up inference.

Estimation of bottom-up proposal function: Given voting results of $g(C, V, D)$, we can estimate bottom-up proposal function. We need three conditional proposal distributions as indicated in Fig. 9.5(b) (dotted arrows). The bottom-up proposal $(q_{boost}(C|\mathbf{e}))$ for object category is simply relative count of evidence votes as Eq. 9.11.

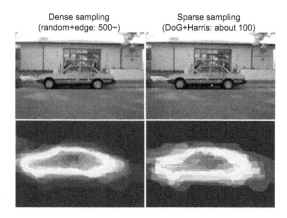

Figure 9.16. The effect of feature sampling method. Dense sample can provide more stable bottom-up inference.

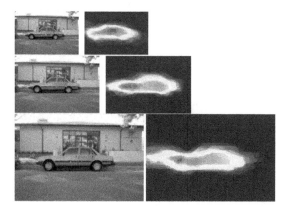

Figure 9.17. Robustness of bottom-up boosting for scale changed test set.

$$q_{boost}(C|\mathbf{e}) = \frac{No.\ of\ votes\ to\ C}{Total\ No.\ of\ votes} \tag{9.11}$$

Given category label C, the viewpoint distribution ($q_{boost}(V|C,\mathbf{e})$) is estimated directly from mean-shift clustering for a set of viewpoints belonging to category C [12].

$$q_{boost}(V(\chi,s)|C,\mathbf{e}) = \sum_m \left(\pi_m N_{\chi,m}(\chi;\mu_\chi,\sigma_\chi^2) \cdot N_{s,m}(s;\mu_s,\sigma_s^2) \right) \tag{9.12}$$

Finally, given object category and viewpoint, we assume thatn the proposal distribution ($q_{boost}(M|C,V,\mathbf{e})$) of figure-ground mask is uniform as Eq. 9.13. This means that an instance of mask M is obtained by randomly thresholding (γ) the voting values of figure-ground masks. The voting values are normalized by the maximal vote so γ is in the range of $[0\ 1]$. Fig. 9.18 shows an examples of boosted distribution of viewpoint and figure-ground mask. For a car category, viewpoint clusters are built using mean-shift and parameterized by GMM. For each viewpoint mode, corresponding proposal distribution of mask is shown below. We can generate figure-ground mask samples by randomly selected threshold (γ).

$$q_{boost}(M|C,V,\mathbf{e}) = 1 \tag{9.13}$$

9.5.2 Top-down inference using multi-modal MCMC sampling

The performance of MCMC-based inference depends on sampling method. In this section, we propose a multi-modal MCMC-sampling for fast and accuate inference. The multi-modal proposal functions are defined as Eq. 9.14 using prior distributions learned from training data such as Eq. 9.5, and boosted proposal distributions in Eq. 9.11,9.12,9.13. The β_i is mixing probability for each random variable sampling. We usually set them as 0.5.

$$\begin{aligned}
q(H \to H'|G,D) &= q_C(C|G,D)q_V(V|C,G,D)q_M(M|C,V,G,D) \\
q_C(C|G,D) &= \beta_1 p(C|D) + (1-\beta_1)q_{boost}(C|G,D) \\
q_V(V|C,G,D) &= \beta_2 p(V|C,D) + (1-\beta_2)q_{boost}(V|C,G,D) \\
q_M(M|C,V,G,D) &= \beta_3 p(M|C,V,D) + (1-\beta_3)q_{boost}(M|C,V,G,D)
\end{aligned} \tag{9.14}$$

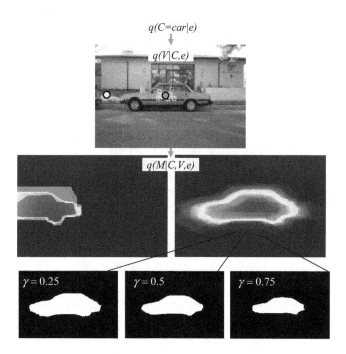

Figure 9.18. Example of bottom-up proposal distribution (see the text for details).

Through conditional sampling from multi-modal distributions, we can generate a hypothesis H' as shown in Fig. 9.12 (right box). Then we can calculate the likehood using Eq. 9.3, 9.4. Hypothesis (H') is accepted with probability α in Eq. 9.7. After convergence, we can get optimal inference result by expectation of accepted samples.

9.6 Experimental results

9.6.1 Performance evaluation

In first experiment, we compare two inference methods for simultaneous object categorization and segmentation: bottom-up only and bottom-up+top-down. We use ROC (receiver operating charateristic) curve as a performance measure [2]. ROC is defined as Eq. 9.15. We use the Caltech Car side dataset for the evaluation [http://www.vision.caltech.edu/Image_Datasets/Caltech101/Caltech101.html]. Randomly selected 15 foreground and background images are used to learn our inference system. Fig. 9.19(top) shows partial examples. In background image, we extract features only background regions. We test 123 cluttered car images as foreground and 123 Google images as background.

$$Positive\ detection\ rate = \frac{Number\ of\ correct positives}{Total\ number\ of\ positives\ in\ data\ set}$$
$$False\ detection rate = \frac{Number\ of\ false\ positives}{Total\ number\ of\ negatives\ in\ data\ set} \quad (9.15)$$

It is important to define control threshold for the correct ROC curve generation. Since our research goal is to categorize and figure-ground segmentation simultaneously, we use one control parameter and two thresholds. As shown in Fig. 9.18, mean-shift clustering (window radius 30) can provide clustered viewpoints. We use this number (k) as a main control parameter. We define an inference is correct positive if $k > k_{th}$, the viewpoint center error is below 50 pixels and the region overlap error ($1 - (R_E \cap R_T)/(R_E \cup R_T)$) is below 30% where R_E is region of estimation and R_T is ground truth region (see Fig. 9.20). In bottom-up with top-down method, we use the same control parameter with additional likelihood ratio test $p(G|O)/P(G|B)$.

Based on such setting, we apply 123 images for positives set and 123 images for negative set as shown in Fig. 9.19 (bottom). By controlling the threshold k_{th}

Figure 9.19. Partial training and test examples of foreground and background data set for car category.

Figure 9.20. Geometric thoreshold for correct positive detection.

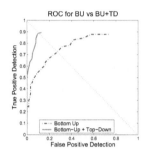

Figure 9.21. ROC curve for car category using Bottom-up only and Bottom-up with top-down method.

Table 9.1. Summary of EER performance for car category detection.

	Ours	[115]	[22]
car side	89.0%	87.3%	88.0%
EER criteria	label+region	label only	label only

from 0 to 100, we can get ROC curve like Fig. 9.21. The equal error rate (EER) for bottom-up only is 73% and that for bottom-up with top-down method is 89%. At this EER, k_{th} is 8. Table 9.1 summarize EER results compared to other related methods. Our EER is higher than others. Furthermore, our system can categorize and segment figure-ground. Fig. 9.22 shows partial car detection results.

As a next evaluation, we check the performance for object occlusion. For this test, we randomly select 50 test images and add artificial squares with size from 20 to 100 pixels at random position. The length of car is 170 pixels in average. We use the parameters at EER. Fig. 9.23 represents the evaluation results. Note that our system is relatively robust to occlusion. Fig. 9.24 shows partial successful object categorization and figure-ground segmentation results. Our system can predict the shape for the occluded regions (see bottom in Fig. 9.24).

We also evaluate our system for Caltech face data set [http://www.robots.ox.ac.uk/~vgg/data3.html]. The face DB consist of 435 faces with clutter and 468 background images. Training is conducted using only 15 random selections. 200 novel face images

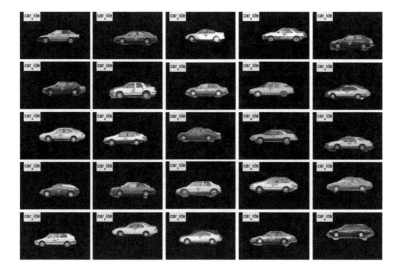

Figure 9.22. Partial examples of correct inference for car category. Note the viewpoint (cross) and figure-ground segmentation.

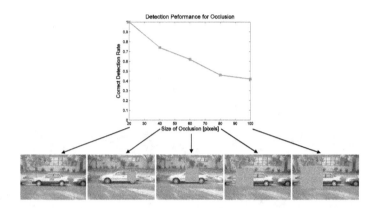

Figure 9.23. Performance for object occlusion. Our system can detect half occluded objects with 50%.

and 200 novel background images are used to check EER. We use the parameters selected in EER for car detection. Table 9.2 summarizes the composition of training set and EER performance. Unsupervised learning requires very large number of training data to provide comparable performance of ours [22, 114]. Partially segmented set can reduce the number of unsegmented training data [93]. Our system relies on fully segmented small training set which provides better performance. Fig. 9.25 shows partial examples of face categorization and figure-ground segmentation results. Note than our system can detect faces very robustly for the face expression and background.

9.6.2 Simultaneous categorization and segmentation

Until now, we evaluate our system in terms of detection performance since almost all the available methods check categorization and localization in ROC performance. Basically, our system is designed for categorization of different categories and figure-ground segmentation. In this experiment, we select five categories from the Caltech-101 DB such as car_side, motorbikes, stop_sign, cup, and faces. Segmented 15 examples per category are used for training and unlearned 15 images per category are used for test. As a baseline method, we use the bag of keypoints which

Figure 9.24. Examples for successful object categorization and segmentation for occluded test set.

Table 9.2. Composition of training set and EER for face test set

Method	No. train (unseg)	No. train (seg)	EER (region error <25%)
[114]	200	0	94.0 %
[22]	220	0	96.4%
[93]	50	10	96.5%
Ours	0	15	97.3%

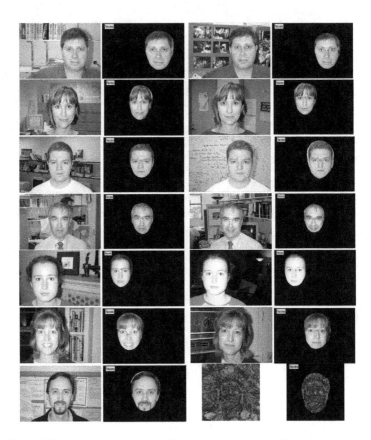

Figure 9.25. Examples of face detection and segmentation for Caltech face set and an ambiguous face.

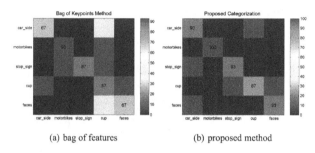

(a) bag of features   (b) proposed method

Figure 9.26. The improvement of categorization by the proposed method.

Figure 9.27. Object categorization and figure-ground segmentation for real test set.

does not use figure-ground information. Fig. 9.26 shows the confusion matrices for the categorization. We use the NNC with KL-divergence for the bag of keypoints. We can get improvement of categorization from 80.0% to 93.3% using our proposed method. Fig. 9.27 shows real test results using real images captured in KAIST.

9.6.3   Cooperative relationship between identification and categorization

Until now, we presented object identification and categorization in real world. Both problems are not independent but strongly correlated. Determining similarity measure in categorization is very difficult. Examplars in identification can provide the required similarity as discussed in this chapter. Likewise, handling novel objects in identification is very difficult. Object categorization can do this job. Fig. 9.28

Generalfframeworkfforftheftcooperativefidentificationfandfcategorization

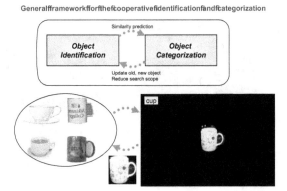

Figure 9.28. Object identification and categorization have cooperative relationship.

shows such cooperative relation for cup examples.

9.7   Summary

In this chapter we proposed an integrated method for object categorization and
figure-ground segmentation for unknown novel objects. Simultaneous categorization and segmentation is very difficult problem. We solve it by utilizing part-part
context, part-object context, and object-background context. Part-part context can
remove or reduce the effect of outliers and part-object context can predict category label, viewpoint with figure-ground mask. By accumulating weak classifier,
we can boost the bottom-up inference. For the top-down inference, we propose a
multi-modal MCMC sampling. Samples are selected in multi-modal distribution
composed of prior term and bottom-up proposal term. This method is converged to
the almost global solution. Through various evaluations, we can conclude that our
integrated system is very useful to object categorization and figure-ground segmentation problem.

CHAPTER X
CONCLUSION

In this dissertation, we focused on object recognition in real world and proposed context-based methods to solve the scalability and generality problem under background clutter and novel objects that degrade object recognition performance.

In the first part, we focused on the scalability issues of object identification in real environments. The object identification means to recognize the same objects learned in off-line. For the scalability, many objects should be discriminated with scalable memory and processing time. Visual contexts in object identification provide very strong discrimination capability. The lowest level of visual context is the pixel context which represents visual information by pixel array. We encoded the pixel context based on the properties human visual systems. An object is decomposed into convex parts and corner parts. Pixels belonging to the part are encoded by the localized receptive field histogram of edge magnitude, orientation, and hue. We call such local feature as G-RIF and it showed high identification (labeling) rate for segmented objects. At the same time, it showed robustness to geometric variations such as scale, view angle, planar rotation, occlusion and to photometric variation of illumination intensity. However, the object labeling using G-RIF showed poor performance due to background clutters in real environment. Part context can discriminate figural region and background region since part context provides grouping property to parts within an object. We modeled the part context by the weight aggregation. Weight is calculated by the similarity of same object label and the proximity of part distance. Several iteration of weight aggregation provided stronger weights to the figural parts and weaker weights to the background parts. The weight aggregated voting showed upgraded object labeling performance for cluttered DB such as KAIST-104 DB and CMU DB. The object identification (simultaneous labeling and localization) is solved by the part-object context. Recognized parts can predict object region. Direct storing part-object relations of many 3D objects is inefficient. For the scalable object representation, we proposed the sharing concept to the parts and the wholes (views). Similar parts are grouped as visual dictionary and similar views in similarity transformation space are grouped as a single view. The scalable object representation and learning method showed scalability in COIL-100 DB and our 112 objects. In static scenes, ambiguous objects cannot be discriminated without higher context. Scene context such as place information can solve this problem. Conversely, ambiguous places can be discriminated by the object context. The whole spatial and hierarchical context were integrated by the hierarchical graphical model (HGM). Piecewise learning and Monte Carlo-based inference provided successful scene interpretation results in building environments. We extended this work by incorporating temporal context in video and showed the practicality.

On the other hand, a few methods were proposed to solve the generality issues. There can be unknown objects in real environments. It usually fails to recognize them with the identification method. The first method utilized the pixel context to represent object categories. Man-made objects show high intra-class variations due to surface markings. We proposed the entropy-guided codebook in bag of visual words method. High entropy codebook can remove surface markings statistically. It showed improved object categorization capability for the Caltech-101 DB. However, pixel context-based object categorization fails to categorize objects in cluttered environment. The part-part context and part-object context can categorize and segment novel objects in clutter. We proposed a directed graphical model to solve such problem. The inference was conducted using the boosted MCMC method. Evaluation showed the generalization capability for novel objects in clutter.

The main contributions of this dissertation are as follows. First, the scalability problem in object recognition is solved by utilizing different level of context. G-RIF based on pixel context showed robust object labeling for segmented objects. Part context provided robust object labeling in background clutter. Part-object context was utilized to label and localized objects with scalable object representation. The spatial and hierarchical contexts were integrated with HGM and showed robust scene interpretation. Static and temporal context showed upgraded performance in video interpretation. Second, the generality problem is solved by utilizing visual context to object categorization. Entropy-guided codebook in pixel context can remove the effect of surface markings. Part-part and part-object context can categorize and segment unknown objects simultaneously in cluttered environment. Finally, cooperative properties between the object identification and categorization is utilized in object recognition problem. Identification can get the generality from categorization and categorization can get the similarity information from instances in identification.

References

[1] A. Agarwal and B. Triggs. Hyperfeatures - multilevel local coding for visual recognition. In *European Conference on Computer Vision*, pages 30–43, 2006.

[2] S. Agarwal and D. Roth. Learning a sparse representation for object detection. In *European Conference on Computer Vision*, pages 113–130, 2002.

[3] S. Arya, D.M. Mount, N.S. Netanyahu, R. Silverman, and A.Y. Wu. An optimal algorithm for approximate nearest neighbor searching fixed dimensions. *Journal of the ACM*, 45(6):891–923, 1998.

[4] M. Bar. Visual objects in context. *Nature Reviews: Neuroscience*, 5:617–629, August 2004.

[5] M. Bar and S. Ullman. Spatial context in recognition. *Perception*, 25(3):324–352, 1996.

[6] S. Belongie, J. Malik, and J. Puzicha. Shape matching and object recognition using shape contexts. *IEEE Trans. Pattern Analysis and Machine Intelligence*, 24(24):509–522, 2002.

[7] A.C. Berg, T.L. Berg, and J. Malik. Shape matching and object recognition using low distortion correspondences. In *IEEE Computer Vision and Pattern Recognition or CVPR*, pages 26–33, 2005.

[8] I. Biederman. Recognition-by-component: A theory of human image understanding. *Psychological Review*, 94:115–147, 1987.

[9] G.M. Boynton. Adaptation and attentional selection. *Nature Neurosci.*, 7(1):8–10, 2004.

[10] P. Carbonetto, N. de Freitas, and K. Barnard. A statistical model for general contextual object recognition. In *European Conference on Computer Vision*, pages 350–362, 2004.

[11] O. Chomat, Vincent Colin de Verdière, D. Hall, and J. L. Crowley. Local scale selection for gaussian based description techniques. In *European Conference on Computer Vision*, pages 117–133, 2000.

[12] D. Comaniciu and P. Meer. Mean shift: A robust approach toward feature space analysis. *IEEE Trans. Pattern Analysis and Machine Intelligence*, 24(5):603–619, 2002.

[13] G. Csurka, C.R. Dance, L. Fan, J. Willamowski, and C. Bray. Visual categorization with bags of keypoints. In *ECCV Workshop on Statistical Learning in Computer Vision*, 2004.

[14] G. Dorkó and C. Schmid. Selection of scale-invariant parts for object class recognition. In *International Conference on Computer Vision*, pages 634–640, 2003.

[15] R.O. Duda, P.E. Hart, and D.G. Stork. *Pattern Classification*. Wiley-Interscience Publication, 2000.

[16] S. Dumais, J. Platt, D. Heckerman, and M. Sahami. Inductive learning algorithms and representations for text categorization. In *Proceedings of the seventh international conference on Information and knowledge management (CIKM'98)*, pages 148–155, 1998.

[17] S. Edelman and H.H. Bulthoff. Orientation dependence in the recognition of familiar and novel views of 3D objects. *Vision Research*, 32:2385–2400, 1992.

[18] B. Epshtein and S. Ullman. Identifying semantically equivalent object fragments. In *IEEE Computer Vision and Pattern Recognition or CVPR*, pages 2–9, 2005.

[19] O.D. Faugeras and M. Hebert. The representation, recognition, and locating of 3-D objects. *IJRR*, 5(3):27–52, 1986.

[20] P. Fearnhead and P. Clifford. On-line inference for hidden markov models via particle filters. *J. R. Statist. Soc. B*, 65:887–899, 2003.

[21] L. Fei-Fei and P. Perona. A bayesian hierarchical model for learning natural scene categories. In *IEEE Computer Vision and Pattern Recognition or CVPR*, pages 524–531, 2005.

[22] R. Fergus, P. Perona, and A. Zisserman. Object class recognition by un-supervised scale-invariant learning. In *IEEE Computer Vision and Pattern Recognition or CVPR*, pages 264–271, 2003.

[23] M. Fink and P. Perona. Mutual boosting for contextual inference. In *Advances in Neural Information Processing Systems*, 2003.

[24] W.T. Freeman, E.C. Pasztor, and O.T. Carmichael. Learning low-level vision. *International Journal of Computer Vision*, 40(1):25–47, 2000.

[25] T.J. Gawne and J.M. Martin. Responses of primate visual cortical V4 neurons to simultaneously presented stimuli. *J. Neurophysiol.*, 88:1128–1135, 2002.

[26] B. Georgescu, I. Shimshoni, and P. Meer. Mean shift based clustering in high dimensions: A texture classification example. In *International Conference on Computer Vision*, volume 1, page 456, 2003.

[27] W.R. Gilks. *Markov Chain Monte Carlo in Practice*. Chapman & Hall/CRC, December 1995.

[28] E. Hayman, B. Caputo, M. Fritz, and J.-O. Eklundh. On the significance of real-world conditions for material classification. In *European Conference on Computer Vision*, pages 253–266, 2004.

[29] X. He, R.S. Zemel, and M.A. Carreira-Perpinan. Multiscale conditional random fields for image labeling. In *IEEE Computer Vision and Pattern Recognition or CVPR*, volume 02, pages 695–702, 2004.

[30] D.P. Huttenlocher and S. Ullman. Object recognition using alignment. *ICCV*, 87:102–111.

[31] K. Jafari-Khouzani and H. Soltanian-Zadeh. Radon transform orientation estimation for rotation invariant texture analysis. *IEEE Trans. Pattern Analysis and Machine Intelligence*, 27(6):1004–1008, 2005.

[32] M.I. Jordan, editor. *Learning in graphical models*. MIT Press, Cambridge, MA, USA, 1999.

[33] Y. Ke and R. Sukthankar. PCA-SIFT: A more distinctive representation for local image descriptors. In *IEEE Computer Vision and Pattern Recognition or CVPR*, pages 506–513, 2004.

[34] Z. Khan, T. Balch, and F. Dellaert. MCMC-based particle filtering for tracking a variable number of interacting targets. *IEEE Trans. Pattern Analysis and Machine Intelligence*, 27(11):1805–1918, 2005.

[35] S. Kim and I.-S. Kweon. Biologically motivated perceptual feature: Generalized robust invariant feature. In *Asian Conference on Computer Vision*, pages 305–314, 2006.

[36] S. Kim and I.-S Kweon. Simultaneous classification and visualword selection using entropy-based minimum description length. In *International Conference on Pattern Recognition*, pages 650–653, 2006.

[37] S. Kim, K.-J. Yoon, and I. S. Kweon. Background robust object labeling by voting of weight-aggregated local features. In *Proceedings of the 18th International Conference on Pattern Recognition (ICPR '06)*, pages 219–222, 2006.

[38] J. Kosecka and F. Li. Vision based topological markov localization. In *Proc. of IEEE International Conference on Robotics and Automation (ICRA)*, 2005.

[39] M. Kouh and M. Riesenhuber. Investigating shape representation in area V4 with hmax: Orientation and grating selectivities. Technical Report AIM2003-021, Massachusetts Institute of Technology, 2003.

[40] S. Kumar and M. Hebert. Multiclass discriminative fields for part-based object detection. In *Snobird Learning Workshop*, Utah, USA, 2004.

[41] S. Kumar and M. Hebert. A hierarchical field framework for unified context-based classification. In *International Conference on Computer Vision*, pages 1284–1291, 2005.

[42] L.J. Latecki and R. Lakämper. Convexity rule for shape decomposition based on discrete contour evolution. *Computer Vision and Image Understanding*, 73(3):441–454, 1999.

[43] S. Lazebnik, C. Schmid, and J. Ponce. A maximum entropy framework for part-based texture and object recognition. In *International Conference on Computer Vision*, pages 832–838, 2005.

[44] S. Lazebnik, C. Schmid, and J. Ponce. Beyond bags of features: Spatial pyramid matching for recognizing natural scene categories. In *IEEE Computer Vision and Pattern Recognition or CVPR*, pages 2169–2178, 2006.

[45] T.S. Lee. Computations in the early visual cortex. *J. Physiology (Paris) 97*, pages 121–139, 2003.

[46] T.S. Lee and D. Mumford. Hierarchical bayesian inference in the visual cortex. *Journal of Optical Society of America, A.*, 20(7):1434–1448, July 2003.

[47] B. Leibe, A. Leonardis, and B. Schiele. Combined object categorization and segmentation with an implicit shape model. In *Workshop on Stat. Learn. in Comp. Vis.*, 2004.

[48] B. Leibe and B. Schiele. Analyzing appearance and contour based methods for object categorization. In *IEEE Computer Vision and Pattern Recognition or CVPR*, volume 2, pages 409–415, 2003.

[49] S.Z. Li. *Markov Random Field Modeling in Image Analysis (Computer Science Workbench)*. Springer-Verlag Telos, June 2001.

[50] Z. Lin, S. Kim, and I.S. Kweon. Recognition-based indoor topological navigation using robust invariant features. In *IEEE/RSJ International Conference on Intelligent Robots and Systems (IROS'05)*, 2005.

[51] T. Lindeberg. Detecting salient blob-like image structures and their scales with a scale-space primal sketch: a method for focus-of-sttention. *International Journal of Computer Vision*, 11(3):283–318, 1993.

[52] D.G. Lowe. Three-dimensional object recognition from single two-dimensional images. *Artificial Intelligence*, 31(3):355–395, 1987.

[53] D.G. Lowe. Local feature view clustering for 3D object recognition. In *IEEE Computer Vision and Pattern Recognition or CVPR*, pages 682–688, 2001.

[54] D.G. Lowe. Distinctive image features from scale-invariant keypoints. *International Journal of Computer Vision*, 60(2):91–110, 2004.

[55] G. Loy and A. Zelinsky. Fast radial symmetry transform for detecting points of interest. *IEEE Trans. Pattern Analysis and Machine Intelligence*, 25(8):959–973, 2003.

[56] J.V. Lucas and W.V. Piet. Estimators for orientation and anisotropy in digitized images. In *1st Conference on Advanced School for Computing and Imaging (ASCI)*, pages 442–450, 1995.

[57] S. Mahamud and M. Hebert. The optimal distance measure for object detection. In *IEEE Computer Vision and Pattern Recognition or CVPR*, 2003.

[58] D. Marr. *Vision: A Computational Investigation into the Human Representation and Processing of Visual Information.* Henry Holt and Co., Inc., New York, NY, USA, 1982.

[59] D. Marr and H.-K. Nishihara. Representation and recognition of the spatial organization of three-dimensional shapes. *Royal Society of London Proceedings Series B*, 200:269–294, February 1978.

[60] K. Mikolajczyk, B.Leibe, and B. Schiele. Local features for object class recognition. In *International Conference on Computer Vision*, pages 1792–1799, 2005.

[61] K. Mikolajczyk, B. Leibe, and B. Schiele. Multiple object class detection with a generative model. In *IEEE Computer Vision and Pattern Recognition or CVPR*, pages 26–36, 2006.

[62] K. Mikolajczyk and C. Schmid. Scale & affine invariant interest point detectors. *International Journal of Computer Vision*, 60(1):63–86, 2004.

[63] K. Mikolajczyk and C. Schmid. A performance evaluation of local descriptors. *IEEE Trans. Pattern Analysis and Machine Intelligence*, 27(10):1615–1630, 2005.

[64] P. Moreels, M. Maire, and P. Perona. Recognition by probabilistic hypothesis construction. In *European Conference on Computer Vision*, pages 55–68, 2004.

[65] B.J. Mundy and A. Zisserman. *Geometric Invariance in Computer Vision.* MIT Press, 1992.

[66] H. Murase and S. Nayar. Visual learning and recognition of 3-D objects from appearance. *International Journal of Computer Vision*, 14:5–24, 1995.

[67] K. Murphy, A. Torralba, and W.T. Freeman. Using the forest to see the trees: A graphical model for recognizing scenes and objects. In *Advances in Neural Information Processing Systems 16*. 2004.

[68] E. Murphy-Chutorian and J. Triesch. Shared features for scalable appearance-based object recognition. In *Proceedings of the Seventh IEEE Workshops on Application of Computer Vision*, volume 1, pages 16–21, 2005.

[69] J. Mutch and D.G. Lowe. Multiclass object recognition with sparse, localized features. In *IEEE Computer Vision and Pattern Recognition or CVPR*, pages 11–18, 2006.

[70] S.K. Nayar, S.A. Nene, and H. Murase. Real-time 100 object recognition system. *IEEE Trans. Pattern Analysis and Machine Intelligence*, 18:1186–1198, 1996.

[71] S.A. Nene, S.K. Nayar, and H. Murase. Columbia Object Image Library (COIL-100). Technical report, Feb 1996.

[72] H. Niemann, G.F. Sagerer, S. Schroder, and F. Kummert. ERNEST: A semantic network system for pattern understanding. *IEEE Trans. Pattern Analysis and Machine Intelligence*, 12(9):883–905, 1990.

[73] D. Nister and H. Stewenius. Scalable recognition with a vocabulary tree. In *IEEE Computer Vision and Pattern Recognition or CVPR*, pages 2161–2168, 2006.

[74] S. Obdrazalek and J. Matas. Sub-linear indexing for large scale object recognition. In WF Clocksin, AW Fitzgibbon, and PHS Torr, editors, *Proceedings of the 16th British Machine Vision Conference*, volume 1, pages 1–10, 2005.

[75] K. Okada and D. Comaniciu. Scale selection for anisotropic scale-space: Application to volumetric tumor characterization. In *IEEE Computer Vision and Pattern Recognition or CVPR*, pages 594–601, 2004.

[76] K. Okuma, A. Taleghani, N. de Freitas, J.J. Little, and D.G. Lowe. A boosted particle filter: Multitarget detection and tracking. In *European Conference on Computer Vision*, pages 28–39, 2004.

[77] A. Opelt, A. Pinz, M. Fussenegger, and P. Auer. Generic object recognition with boosting. *IEEE Trans. Pattern Analysis and Machine Intelligence*, 28(3):416–431, 2006.

[78] A. Opelt, A. Pinz, and A. Zisserman. A boundary-fragment-model for object detection. In *European Conference on Computer Vision*, pages 575–588, 2006.

[79] A. Pasupathy and C.E. Connor. Shape representation in area V4: Position-specific tuning for boundary conformation. *J. Neurophysiol*, 86(5):2505–2519, 2001.

[80] M. Reisenhuber and T. Poggio. Models of object recognition. *Nature Neuroscience*, 3:1199–1204, 2000.

[81] D. Reisfeld, H. Wolfson, and Y. Yeshurun. Context-free attentional operators: The generalized symmetry transform. *International Journal of Computer Vision*, 14(2):119–130, 1995.

[82] D.L. Ringach. Spatial structure and symmetry of simple-cell receptive field in macaque primary visual cortex. *J. Neurophysiol.*, 88:455–463, 2001.

[83] B. Ristic, S. Arulampalam, and N. Gordon. *Beyond the Kalman, Filter-Particle Filters for Tracking Applications*. Artech Hous, London, 2004.

[84] K.-S. Roh and I.-S. Kweon. 2-D object recognition using invariant contour descriptor and projective refinement. *Pattern Recognition*, 31(4):441–455, 1998.

[85] C.A. Rothwell. *Object Recognition through Invariant Indexing*. Oxford University Press, 1995.

[86] B. Schiele and J.L. Crowley. Recognition without correspondence using multidimensional receptive field histograms. *International Journal of Computer Vision*, 36(1):31–50, 2000.

[87] C. Schmid. A structured probabilistic model for recognition. In *IEEE Computer Vision and Pattern Recognition or CVPR*, volume II, pages 485–490, June 1999.

[88] C. Schmid, R. Mohr, and C. Bauckhage. Evaluation of interest point detectors. *International Journal of Computer Vision*, 37(2):151–172, 2000.

[89] S. Se, D.G. Lowe, and J.J. Little. Vision-based global localization and mapping for mobile robots. *IEEE Trans. on Robotics*, 21(3):364–375, 2005.

[90] T. Serre, L. Wolf, and T. Poggio. Object recognition with features inspired by visual cortex. In *IEEE Computer Vision and Pattern Recognition or CVPR*, 2005.

[91] A. Shokoufandeh, D. Macrini, S. Dickinson, K. Siddiqi, , and S. W. Zucker. Indexing hierarchical structures using graph spectra. *IEEE Trans. Pattern Analysis and Machine Intelligence*, 27(7):1125–1140, 2005.

[92] A. Shokoufandeh, I. Marsic, and S.J. Dickinson. View-based object matching. In *International Conference on Computer Vision*, pages 588–595, 1998.

[93] J. Shotton, A. Blake, and R. Cipolla. Contour-based learning for object detection. In *International Conference on Computer Vision*, pages 503–510, 2005.

[94] A. Stein and M. Hebert. Incorporating background invariance into feature-based object recognition. In *Workshop on Applications of Computer Vision (WACV'05)*, pages 37–44, 2005.

[95] A.J. Storkey. Dynamic trees: A structured variational method giving efficient propagation rules. In C. Boutilier and M. Goldszmidt, editors, *Uncertainty in Artificial Intelligence*, pages 566–573. Morgan Kauffmann, 2000.

[96] A.J. Storkey and C.K.I. Williams. Image modelling with position-encoding dynamic trees. *IEEE Trans. Pattern Analysis and Machine Intelligence*, 25(7):859–871, 2003.

[97] E.B. Sudderth, A. Torralba, W.T. Freeman, and A.S. Willsky. Learning hierarchical models of scenes, objects, and parts. In *International Conference on Computer Vision*, pages 1331–1338, 2005.

[98] C. Sutton and A. McCallum. Piecewise training of undirected models. In *21st Conference on Uncertainty in Artificial Intelligence*, 2005.

[99] M. Tico and P. Kuosmanen. Fingerprint matching using an orientation-based minutia descriptor. *IEEE Trans. Pattern Analysis and Machine Intelligence*, 25(8):1009–1014, 2003.

[100] S. Todorovic and M.C. Nechyba. Interpretation of complex scenes using generative dynamic-structure models. In *Proceedings of the 2004 Conference on Computer Vision and Pattern Recognition Workshop (CVPRW'04)*, volume 12, page 184, 2004.

[101] A. Torralba, K.P. Murphy, and W.T. Freeman. Sharing features: Efficient boosting procedures for multiclass object detection. In *IEEE Computer Vision and Pattern Recognition or CVPR*, volume 2, 2004.

[102] A. Torralba, K.P. Murphy, W.T. Freeman, and M.A. Rubin. Context-based vision system for place and object recognition. In *International Conference on Computer Vision*, pages 273–280, 2003.

[103] Z. Tu, X. Chen, A.L. Yuille, and S.-C. Zhu. Image parsing: Unifying segmentation, detection, and recognition. *International Journal of Computer Vision*, 63(2):113–140, 2005.

[104] M. Turk and A. Pentland. Eigenfaces for recognition. *Journal of Cognitive Neuroscience*, 3(1):71–86, 1991.

[105] A. Vailaya, M. A. T. Figueiredo, A. K. Jain, and H.-J. Zhang. Image classification for context-based indexing. *IEEE Trans. Image Processing*, 10(1):117–130, 2001.

[106] R. VanRullen. Visual saliency and spike timing in the ventral visual pathway. *J Physiol Paris*, 97(2-3):365–377, 2003.

[107] V.N. Vapnik. *The nature of statistical learning theory*. Springer-Verlag New York, Inc., New York, NY, USA, 1995.

[108] N. Vasconcelos and M. Vasconcelos. Scalable discriminant feature selection for image retrieval and recognition. *IEEE Computer Vision and Pattern Recognition or CVPR*, 02:770–775, 2004.

[109] J. Vermaak, A. Doucet, and P. Pérez. Maintaining multi-modality through mixture tracking. In *International Conference on Computer Vision*, pages 1110–1116, 2003.

[110] P. Viola and M.J. Jones. Robust real-time face detection. *International Journal of Computer Vision*, 57(2):137–154, 2004.

[111] B. Vo, S. Singh, and A. Doucet. Sequential monte carlo methods for bayesian multi-target filtering with random finite sets. *IEEE Trans. Aerospace and Electronic Systems*, 41(4):1224–1245, 2005.

[112] J. Vogel and B. Schiele. Natural scene retrieval based on a semantic modeling step. In *International Conference on Image and Video Retrieval CIVR 2004, Dublin, Ireland, LNCS*, volume 3115, July 2004.

[113] C. Wallraven, B. Caputo, and A. Graf. Recognition with local features: the kernel recipe. In *International Conference on Computer Vision*, pages 257–264, 2003.

[114] M. Weber, M. Welling, and P. Perona. Unsupervised learning of models for recognition. In *European Conference on Computer Vision*, pages 18–32, 2000.

[115] J. Willamowski, D. Arregui, G. Csurka, C. Dance, and L. Fan. Categorizing nine visual classes using local appearance descriptors. In *ICPR 2004 Workshop Learning for Adaptable Visual Systems Cambridge*, 2004.

[116] J. Winn, A. Criminisi, and T. Minka. Object categorization by learned universal visual dictionary. In *International Conference on Computer Vision*, pages 1800–1807, 2005.

[117] J. Winn and J. Shotton. The layout consistent random field for recognizing and segmenting partially occluded objects. In *IEEE Computer Vision and Pattern Recognition or CVPR*, pages 37–44, 2006.

[118] R.H. Wurtz and E.R. Kandel. Perception of motion, depth and form. *Principles of Neural Science*, pages 548–571, 2000.

[119] J.S. Yedidia, W.T. Freeman, and Y. Weiss. Understanding belief propagation and its generalization. *In G. Lakemayer and B. Nebel, editors, Exploring Artificial Intelligence in the New Milennium, Morgan Kauffmann*, pages 509–522, 2002.

[120] S.X. Yu and J. Shi. Object-specific figure-ground segregation. In *IEEE Computer Vision and Pattern Recognition or CVPR*, pages 39–45, 2003.

[121] H. Zhang, A.C. Berg, M. Maire, and J. Malik. SVM-KNN: Discriminative nearest neighbor classification for visual category recognition. In *IEEE Computer Vision and Pattern Recognition or CVPR*, pages 2126–2136, 2006.

[122] L. Zhang, S.Z. Li, Z.Y. Qu, and X. Huang. Boosting local feature based classifiers for face recognition. In *CVPR Workshop on Face Processing in Vieo (FPIV'04)*, page 87, 2004.